ELECTRICITY FOR HEATING, AIR CONDITIONING, AND REFRIGERATION

Electricity for Heating, Air Conditioning, and Refrigeration

Rex Miller
State University College at Buffalo

HARCOURT BRACE JOVANOVICH, PUBLISHERS
Technology Publications

San Diego   New York   Chicago   Austin   Washington, D.C.
London   Sydney   Tokyo   Toronto

Copyright © 1988 by Harcourt Brace Jovanovich, Inc.

All rights reserved. No part of this publication may be reproduced or transmitted in any form or by any means, electronic or mechanical, including photocopy, recording, or any information storage and retrieval system, without the permission in writing from the publisher.

Requests for permission to make copies of any part of the work should be mailed to: Permissions, Harcourt Brace Jovanovich, Publishers, Orlando, Florida 32887.

ISBN: 0-15-520947-7
Library of Congress Catalog Card Number: 87-82957
Printed in the United States of America

This book is dedicated to
Patricia Ann Navara Miller

Preface

This book is intended to provide an introduction to the basic principles of electricity and electronics as they apply to refrigeration, heating, and air-conditioning systems. It combines basic principles, using very little math, with the latest applications found in the HVAC industry.

The fields of refrigeration, air conditioning, and heating are undergoing some very rapid changes. The advent of the computer chip has made it possible to control heating and cooling systems precisely and with a great deal of freedom in programming their applications. The new high-efficiency furnaces utilize the chip both for sequencing and for protection from accidental damage. Most of these programmable controllers provide instructions and technical bulletins. They are numerous, and each has its own approach to solving a given problem. This book will make it possible for you to understand these instructions and improvements.

Future technicians need to deal with the fact that change is inevitable, and that they will have to keep up with the latest developments as long as they work in the field. It is hoped that this book will make that task easier.

ACKNOWLEDGMENTS

No author works without being influenced and aided by others, and this book is no exception. A number of people cooperated in providing technical data and illustrations, and for this I am grateful. I would like to thank those organizations that so generously contributed information and illustrations. The following have been particularly helpful:

Amprobe Instrument Division of SOS Consolidated, Inc.

Carrier Corporation

Eveready Division of Union Carbide

General Controls

General Electric Company

Honeywell, Inc.

Kelvinator Appliance Company

Lennox Industries (Toronto, Canada)

National Safety Council

Robert Shaw Controls Company

Sporlan Valve Company

Tecumseh Products Company

Tyler Refrigeration Corporation

Wadsworth Electric Company

Wagner Electric Company

Westinghouse Electric Company

Weston Electrical Instruments Company

REX MILLER

Contents

PREFACE vii

CHAPTER 1
INTRODUCTION TO ELECTRICITY 3

Static Electricity and Magnetism, 3   Electricity, 5   Electricity's Future, 5   Matter and Electricity, 6   Matter and Molecules, 6   Solids, Gases, and Liquids, 6   Elements and Compounds, 7   Molecule, 7   Mixture, 7   Atom, 8   Atoms, Electrons, Protons, and Neutrons, 8   Properties of Electrons, 9   Orbiting Electrons, 11   Electrical Charge, 11   Neutron, 12   Outer Shell, 12   Valence Electrons and Ions, 12   A Practical Unit for Charges, 15   The Volt, 15   Controlling Electrons, 15   Difference of Potential (Voltage), 15   Electron Flow (Current), 16   Conductors, 16   Resistance, 17   The Electric Circuit, 17   Connecting a Circuit, 19   Switches Control Electron Flow, 19   Schematic, 20   Review Questions, 20

CHAPTER 2
CURRENT, VOLTAGE, RESISTANCE, POWER, AND OHM'S LAW 23

Sources of Electricity, 23   Units of Measurement, 25   Ohm, 26   Prefixes, 27   Siemens, 28   Ohm's Law, 28   Power, 35   Joule, 35   Watt, 36   Putting Electricity to Work, 36   Resistors and Heat, 36   Mechanical Energy, 37   Kilowatt-Hour, 38   Review Questions, 39

CHAPTER 3
RESISTORS, COLOR CODE, COMPONENTS, AND SYMBOLS 43

Resistors, 43 Color Code, 44 Variable Resistors, 47 Types of Resistors, 49 Types of Capacitors, 53 Types of Inductors, 55 Transformers, 57 Semiconductors, 58 Switches, 59 Relays, 59 Fuses and Circuit Breakers, 60 Lamps, 62 Meters, 63 Review Questions, 64

CHAPTER 4
SERIES AND PARALLEL CIRCUITS 67

Series Circuit, 67 Parallel Circuits, 72 Series-Parallel Circuits, 74 Review Questions, 83

CHAPTER 5
MAGNETISM, SOLENOIDS, AND RELAYS 85

Permanent Magnets, 85 Temporary Magnets, 86 Electromagnets, 87 Magnetic Theory, 87 Electromagnetism, 91 Electromagnets, 92 The Solenoid, 94 Applications, 99 Transformers for Low-Voltage Controls, 103 Review Questions, 104

CHAPTER 6
ELECTRICAL MEASURING INSTRUMENTS 107

Types of Meter Movements, 107 Parallax Error, 112 Ammeter, 112 Voltmeter, 115 AC Ammeter, 116 AC Voltmeter, 116 Ohmmeters, 116 Multimeter, 118 Megger, 118 Digital Meter, 119 Other Instruments, 119 Using an Ohmmeter, 122 Review Questions, 124

CHAPTER 7
ELECTRICAL POWER: DIRECT CURRENT AND ALTERNATING CURRENT 127

Types of Batteries, 127 Dry Cells, 128 Battery Specifications, 128 Connecting Cells, 132 Battery Maintenance, 134 Nickel-Cadmium Cells, 135 Alkaline Cell, 135 Alternating Current, 135 Sine Wave, 136 Sine-Wave Characteristics, 137 Polyphase Alternating Current, 139 Three-Phase Connections, 141 Electrical Properties of Delta and Wye, 141 Review Questions, 144

CHAPTER 8
INDUCTORS AND TRANSFORMERS   147

Inductors, 147   Changing Inductance, 148   Self-Inductance, 148   Mutual Inductance, 151   Inductive Reactance, 152   Power in an Inductive Circuit, 152   Uses for Inductive Reactance, 153   Transformers, 154   Voltage Transfer, 156   Power Transformers, 158   Audio Frequency and Radio Frequency Transformers, 158   Autotransformers, 159   Transformer Losses, 161   Inductive Circuits, 162   Utilizing the Inductive Delay, 164   Review Questions, 164

CHAPTER 9
CAPACITORS AND CAPACITIVE REACTANCE   167

The Capacitor, 168   How the Capacitor Works, 169   Capacity of a Capacitor, 170   Breakdown Voltage, 171   Basic Units of Capacitance, 171   Types of Capacitors, 171   Electrolytic Capacitors, 173   Working Voltage, Direct Current (WVDC), 175   Capacitive Reactance, 175   Capacitor Causes a Lagging Voltage, 177   Checking Capacitors, 177   Review Questions, 179

CHAPTER 10
SINGLE-PHASE AND THREE-PHASE ALTERNATING CURRENT   181

Resistance, Capacitance, and Inductance, 181   Power, 182   Power Factor, 184   Distributing Electric Power, 185   Polyphase, 185   Circuit Breakers, 190   Review Questions, 191

CHAPTER 11
SOLID-STATE CONTROLS   193

Semiconductor Principles, 193   Diode, 194   Silicon-Controlled Rectifiers, 197   Transistors, 197   Integrated Circuits, 200   Thermistor Sensing, 202   Humidity Sensing, 203   Controllers, 208   Electronic Controllers, 208   Differential Amplifiers, 208   Actuators, 210   Other Devices, 211   Solid-State Compressor Motor Protection, 211   Review Questions, 217

CHAPTER 12
ALTERNATING CURRENT MOTORS   219

Shaded-Pole Motor, 221   Split-Phase Motor, 223   Repulsion Start, Induction Run Motor, 227   Capacitor-Start Motor, 228   Permanent Split-Capacitor (PSC), 232   Capacitor-Start, Capacitor-Run Motor, 236   Three-Phase Motor, 238   Capacitor Ratings, 240   Start Capacitors and Bleeder Resistors, 240   Motor Protectors, 242   Compressor Motor Relays, 243   Review Questions, 245

CHAPTER 13
ELECTRICAL SAFETY 247

Safety Precautions, 247 Main Switches, 248 Portable Electrical Tools, 249 Types of Circuit Protectors, 252 Review Questions, 254

CHAPTER 14
CONTROL DEVICES 257

Power Relay, 257 Time-Delay Relays, 262 Solenoids, 262 Thermostats, 263 Microprocessor Thermostats, 267 Thermostat Adjustments, 270 Limit Switches, 271 Pressure Control Switches, 272 Water Tower Controls, 273 Review Questions, 275

CHAPTER 15
HEATING CIRCUITS 277

Basic Gas Furnace Operation, 278 Basic Electric Heating System, 279 Ladder Diagrams, 280 Manufacturer's Diagrams, 282 Field Wiring, 282 Low-Voltage Wiring, 284 Heat Pumps, 286 High-Efficiency Furnaces, 294 Troubleshooting the Pulse™ Furnace, 302 Review Questions, 304

CHAPTER 16
AIR-CONDITIONING CIRCUITS 307

Basic Air-Conditioning Unit, 307 Schematics, 311 Ladder Diagrams, 319 Troubleshooting, 320 Review Questions, 321

CHAPTER 17
REFRIGERATION CIRCUITS 323

Refrigerator-Freezer Combination, 324 Defrosting, 328 Other Devices, 332 Troubleshooting, 335 Rapid Electrical Diagnosis, 338 Energy-Saver Switch, 339 Review Questions, 342

CHAPTER 18
TROUBLESHOOTING 345

Safety, 345 Compressor Problems, 346 Low-Voltage Operation, 351 Using a System to Troubleshoot (Electrical), 351 Using Meters to Check for Problems, 355 Using a Voltmeter for Troubleshooting Electric Motors, 355 Testing for a Short Circuit Between Run and Start Windings, 360 Capacitor Testing, 360 Troubleshooting Procedure, 363

INDEX 365

ELECTRICITY FOR HEATING, AIR CONDITIONING, AND REFRIGERATION

PERFORMANCE OBJECTIVES

Understand how matter and electricity are related.

Understand how liquids, solids, and gases are similar, yet different.

Understand how atoms, electrons, protons, and neutrons are all related to the production of electricity.

Understand how electricity is put to work in the electric circuit.

Understand how switches control the flow of electricity.

CHAPTER 1

Introduction to Electricity

Electricity is as old as the universe. But knowledge about it is relatively new. Early humans were aware of electricity in the form of lightning. They learned of its power when they saw it start fires and kill people and animals. But it was only about 300 years ago that people began to learn the basic laws of electricity. And only about 120 years have passed since electricity was first put to work. It has been only 100 years since the first practical electric lamp was invented, only 80 years since the vacuum tube was invented, and less than 40 years since the transistor was invented. Despite this brief time, electricity has greatly changed people's lives—our lives.

The universe consists of atoms and every atom contains at least one electron. *An electron is the smallest particle of an atom and has a negative electrical charge.* When the movement of electrons is controlled, they are capable of doing work.

STATIC ELECTRICITY AND MAGNETISM

There are two types of electrical effects: static electricity and magnetism.

The Greek philosopher Thales, who lived about 2500 years ago, is credited with discovering static electricity.

Magnetism is the ability of an object to attract other objects. It was discovered about 2600 B.C., or about 100 years before the discovery of static electricity. It is not certain who discovered magnetism first. Some historians say it was first observed by the Chinese. Others say that it was the Greeks. The discoverer noted that certain heavy stones have the power to attract and lift iron and certain other stones. The material in these stones is called *magnetite*. It was named by the Greeks for the

4 INTRODUCTION TO ELECTRICITY

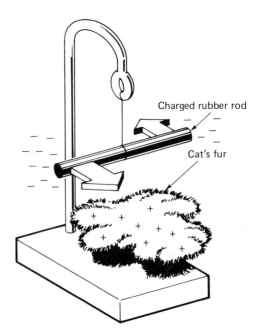

FIGURE 1-1 Unlike charges attract.

province of Magnesia in Asia Minor, where the stones were first found. Today, the power of this stone is called *magnetism*.

These discoveries led to extensive studies of magnetism and static electricity (see Figs. 1-1 and 1-2).

Other people studied electricity and magnetism during the sixteenth century and later in the nineteenth century. Some of these people have electrical terms named for them. You may be familiar with Ampere, Volta, Coulomb, Oersted, Ohm, and Galvani (see Fig. 1-3).

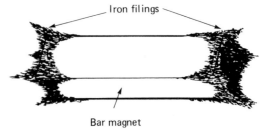

FIGURE 1-2 Iron filings cling to the ends of a permanent magnet. Note the north and south poles.

FIGURE 1-3 Georg Ohm demonstrates his theories to some of his colleagues.

ELECTRICITY

One of the most famous experiments in electricity occurred in 1752. In that year, Benjamin Franklin used a kite and a key to successfully draw lightning from the sky. He was trying to prove that electricity is a fluid. From these and other experiments, Benjamin Franklin is credited with forming the theory of positive and negative charges. It was another 80 years, though, before someone discovered that there is a relationship between magnetism and electron flow.

ELECTRICITY'S FUTURE

The effects of the knowledge and use of electricity were profound. Difficult tasks became easy. Old methods were replaced by new. Electrical machines relieved people of back-breaking labor. Machines could do the job better and cheaper than people. Fears that new inventions and methods would displace workers and create widespread unemployment did not prove true. Instead of a loss of jobs, electricity lead to new industries and new jobs. The new industries required more people than were replaced by machines.

Even today we hear about the possibilities of people losing their jobs because of machines and robots. Automation is the use of machines that are controlled by other machines and devices instead of people. It is another step in technical progress. It makes possible more things faster and better. Automation and robots create more jobs and a need for more skilled people. Trained people are needed to design, build, and maintain electrical equipment.

One of the greatest uses of electricity is in the production of ice and cooling for human comfort. Refrigeration and air conditioning rely exclusively on the ability of electricity to pump a fluid or gas through a system. Electricity is also used to control the temperature in heating, air-conditioning, and refrigeration systems.

MATTER AND ELECTRICITY

The name electricity implies the importance of the almost weightless, invisible part of an atom called an electron. It is electrons that cause electricity. Electricity is defined as the movement of electrons along a conductor.

An electron is only one part of an atom. An atom is only one part of a molecule. None of these can be seen by the unaided eye. Thus, most actions in electrical circuits cannot be seen. An electrical circuit can appear motionless although great activity is happening within it at the atomic level.

The electron can be controlled. Control of the electron is the task of an electrician, electrical engineer, or anyone else working with electricity. Electricity can perform work. It can kill. Using it requires knowledge of such things as matter and mass.

MATTER AND MOLECULES

Matter surrounds us. It is said to be anything that occupies space. Thus, all physical objects are composed of matter.

Matter has mass. Mass is defined as the resistance an object offers to a change in motion. The tighter the matter is packed together, the greater is its mass. Thus, the greater is its resistance to any change in motion.

SOLIDS, GASES, AND LIQUIDS

The three basic forms of matter, shown in Fig. 1-4, are solid, liquid, and gas. A solid, such as a glass container, is stable and self-supporting. By definition, a solid substance is one that offers a large resistance to forces that might change its shape. A liquid, such as water, maintains a definite volume, but assumes the shape of the container in which it is placed. A gas, such as the air we breathe, has no definite volume. It can be expanded or compressed to the shape or size of any container.

The different forms of solid, liquid, and gaseous matter are called substances.

MIXTURE   7

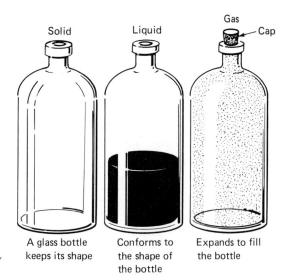

FIGURE 1-4 Behavior of three forms of matter: solid, liquid, and gas.

A glass bottle keeps its shape

Conforms to the shape of the bottle

Expands to fill the bottle

Pure water at room temperature is a liquid substance. All samples of pure water are identical. Pure iron is a solid, and pure carbon dioxide is a gaseous substance.

ELEMENTS AND COMPOUNDS

The two classes of substances are elements and compounds. An element is a pure substance that cannot be divided into any more basic substances by chemical change. The elements (there are more than 100) are the simplest forms of matter. Some examples of elements are hydrogen, oxygen, germanium, silicon, gold, and silver. A compound is a substance composed of two or more elements chemically combined. Water, for example, is a compound made up of the elements hydrogen and oxygen. Common table salt (sodium chloride) is a compound made from the elements sodium and chlorine.

MOLECULE

A molecule is the smallest quantity of any compound that can exist and still retain all the properties of the compound. For example, the elements hydrogen and oxygen, when combined, produce the compound water. A molecule of water, however, is the smallest quantity of water that has all the characteristics of water.

MIXTURE

Elements and compounds can also form mixtures. A mixture is a mixing of two or more substances in which the properties of each substance are not changed. Water

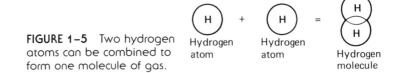

FIGURE 1-5 Two hydrogen atoms can be combined to form one molecule of gas.

is a compound. But salt and water is a mixture because the salt can be separated from the water by simple filtering or evaporation.

ATOM

An atom is the smallest part of an element that retains all the qualities of the element. Atoms are the building blocks from which all substances are made. Figure 1-5 shows two atoms of hydrogen gas combining to form one molecule of hydrogen gas. Some substances (compounds) are formed by the chemical combination of different elemental atoms. Water is one of these (see Fig. 1-6). A molecule of water consists of two hydrogen atoms and one oxygen atom.

An important part of an atom is the electron. An electric current is the result of the controlled movement of electrons in a substance.

ATOMS, ELECTRONS, PROTONS, AND NEUTRONS

The atom is the basic building block of matter, but it can be divided into many particles. The three major particles of the atom are electrons, protons, and neutrons. These three particles are important because they affect the electrical properties of the material. The protons and neutrons form the central mass or nucleus of the atom. One or more electrons circle the nucleus. The simplest atom is the hydrogen atom (see Fig. 1-7A). The nucleus is almost 2000 times heavier than an electron. The electron is the smallest part known. It takes more than 28 billion, billion, billion electrons to weigh 1 ounce.

There is one nucleus in each atom. Elements differ because there are many combinations of orbiting electrons and groupings of protons and neutrons within the nucleus. Each element is made up of atoms having one particular combination of

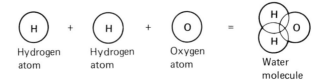

FIGURE 1-6 Two hydrogen atoms can be combined with one oxygen atom to form a molecule of water.

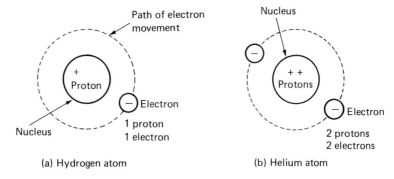

FIGURE 1-7 Simple atoms that contain one and two orbiting electrons. (A) Hydrogen atom. (B) Helium atom.

nucleus and orbital electrons. Each compound substance is made up of a particular arrangement of these atoms.

The simplest atom is the hydrogen atom (see Fig. 1-7A). It contains a nucleus and one electron. A helium atom (Fig. 1-7B) contains a nucleus and two electrons. In other atoms there is more than one shell of orbiting electrons. Copper has four shells. Some electrons have as many as seven electron shells. The fourth shell around the copper nucleus is made up of only one electron. It is easily moved from one atom to the other by heat or magnetism to produce an electron flow or electric current.

The carbon atom and the copper atom are shown in Fig. 1-8. Notice that the carbon atom has only two shells, but has four electrons in the outer shell. The copper atom has only one electron in the outer shell. These two atoms are very important in electricity.

A comparison such as the one shown in Fig. 1-9 is often made between our solar system and an atom. The nucleus of the atom is compared to the sun. Electrons revolving around the nucleus are compared to the planets revolving around the sun. A major difference between the two systems is the orbital paths of the planets and electrons. Figure 1-9 shows this difference. The planets have orbits in a fairly common plane during their trips around the sun. In contrast, the orbits of the electrons around the nucleus are constantly changing (this is present-day theory and subject to change later), and their paths eventually produce spherical shells around the nucleus. The arrangement of these spherical paths of the electrons and direction of their rotation around the nucleus determine the magnetic properties of the substance.

PROPERTIES OF ELECTRONS

The electrical properties of a substance are influenced by the number and arrangement of the electrons in the outermost shell. These electrons, located in the outer

10 INTRODUCTION TO ELECTRICITY

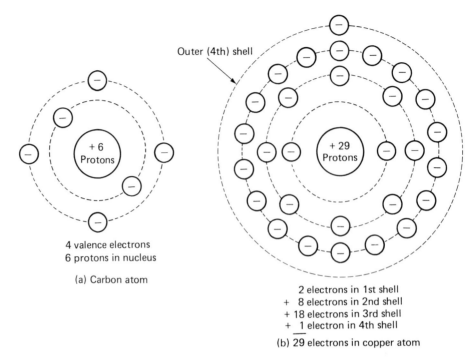

FIGURE 1-8 Some atoms contain more than one shell of orbiting electrons. (A) Carbon atom. (B) Copper Atom.

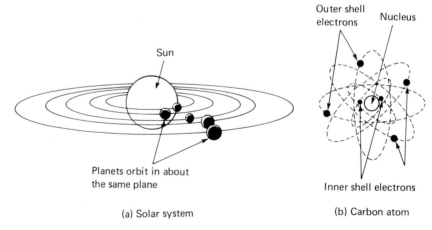

FIGURE 1-9 Comparison of an atom with the solar system. (A) Solar system. (B) Carbon atom.

shell, are called *valence electrons*. Keep in mind that all electrons are alike. They are the same in all atoms. Electrons can be moved among like and unlike atoms. The application of an electrical force causes electrons to move from atom to atom in a controlled manner. The movement of electrons from atom to atom is called *electric current*. Because all electrons are the same, the basic atomic makeup of a substance (such as copper) is not changed by electron movement.

ORBITING ELECTRONS

Orbiting electrons do not leave the atom. Orbiting planets do not leave the solar system. People can orbit the earth and return without being lost in space.

Two forces prevent the electrons from leaving the atom. One is the force or pull of gravity. It is the same pull that keeps things on earth. Energy supplied to a satellite tends to cause it to be pulled away from the earth. There is an attraction between the nucleus and the electron that causes it to be held in orbit around the atom. When a force or energy is applied sufficient to cause it to move from the orbiting path, it moves to the next atom and in moving can produce what we call electricity. This movement can cause work to be done.

ELECTRICAL CHARGE

Figure 1-10A illustrates the electrical charge. A ball is attached to a string and made to swing in a circle. The swinging ball tends to move away from the hand. But it is

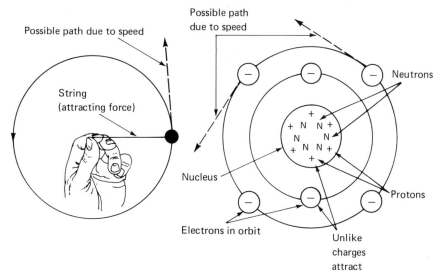

FIGURE 1-10 Comparison of a ball swinging attached to a string and an electron swinging around the nucleus of an atom. (A) Ball on a string. (B) Location of electrons.

held by the string. This is like Fig. 1-10B in which the electrons swinging around an atom are pulled to the center nucleus. The speed of rotation causes them to follow an orbital path around the nucleus. The force between the electron and the nucleus is called an electrical charge.

The electron possesses a *negative* charge. The nucleus has the opposite polarity, a *positive* charge. In the nucleus, however, the positive charge is carried by protons. Thus, for every electron in orbit there is a proton in the nucleus. This is shown in Fig. 1-10B. There the atom has six electrons in orbit and six protons in the nucleus. The simple hydrogen atom has one electron in orbit and one proton in its nucleus.

NEUTRON

Another major part of the atom is the *neutron*. The neutron is also a part of the nucleus. However, it has no charge. Neutrons and protons have nearly the same weight. They determine, for the most part, the mass of the atom. The electron has little weight and little mass.

OUTER SHELL

A basic law of electric charges is that like charges repel and unlike charges attract. The effect of charges on freely moving bodies is shown in Fig. 1-11. In the atom the positive charge of the nucleus (protons) attracts the electrons. However, the speed and energy of the electrons causes them to maintain their orbital paths. Since the forces in the atom are balanced, the electrical charges are balanced. Thus, the atom remains stable and neutral.

Electrons in the outer shell of an atom (called valence electrons) have a higher energy than electrons in the shells closer to the nucleus. External force can add energy to the valence electrons. This added energy permits their escape from the atom. Such *free electrons* can move from atom to atom. Being free, they are used for electric current.

VALENCE ELECTRONS AND IONS

Normally, we are concerned only with the valence electrons (outer shell) because these are the easiest to free. When one (or more) electron is removed from or added to the outer shell of an atom, the atom becomes charged. It is no longer neutral. Then the atom is called an *ion*. When an electron is lost, the atom takes on a positive charge because there are more protons in the nucleus than orbiting electrons. The atom is then called a positive ion. Ionization of air is used in some "air-conditioning" systems.

Sometimes it is possible to add an electron to the outer shell. This results in a

VALENCE ELECTRONS AND IONS 13

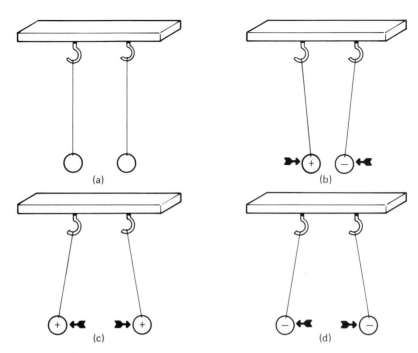

FIGURE 1-11 Unlike charges attract each other. Like charges repel each other. (A) No charge means there is no attraction or repulsion. (B) Positive and negative charges are attracted to each other (unlikes attract). (C) Positive charges repel each other; likes repel. (D) Negative charges repel each other; likes repel.

charged atom that is a negative ion. It is negative because there are more orbiting electrons than protons in its nucleus.

Unlike charges attract. Therefore, a positive ion will attract an electron or any negatively charged body. A negative ion, however, will repel an electron or any negatively charged body.

Unlike charges on two bodies mean there is a difference between them (see Fig. 1-12). A difference in charge exists between the four pairs of charged bodies in Fig. 1-13. This difference is 4. If a conducting path for electrons is made between any pair of bodies in Fig. 1-13, the same number of electrons would have to move from left to right in the illustration to neutralize the charges. When two bodies have the same charge and same polarity, there is no difference between them.

So far we have considered the charge in terms of electrons or numbers. Now we need to give the charge a name. A name is also needed for the difference that exists between these charges.

Any substance, molecule, or atom may have a negative or positive charge. Or it may be neutral. How much *more negative* or *more positive* can one body be charged relative to another? If a comparison is made, some unit of measurement

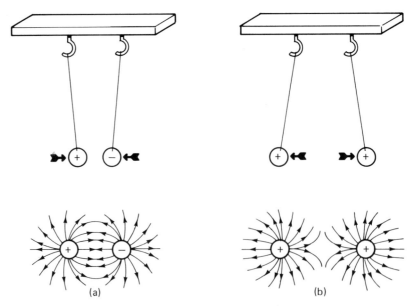

FIGURE 1-12 Invisible force fields extend outward from charged particles. (A) Lines of force unite and draw the unlike charges together. (B) Lines of force do not unite, so they repel.

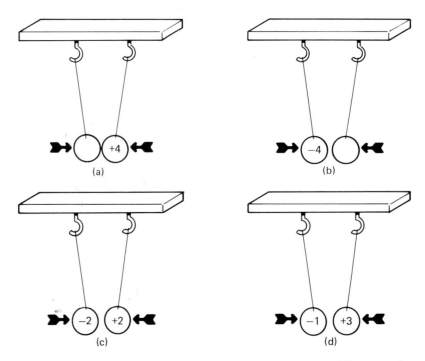

FIGURE 1-13 Four pairs of charged bodies with the same difference of potential. Each pair of charges is attracted by the same amount of force.

must be used. Some standard reference should be used as a basic unit of measurement.

The smallest negative charge is already understood to be that of the electron. And the charge of a proton is the smallest positive electric charge. Such charges are too small and not useful in terms of establishing a basic unit of measurement.

A PRACTICAL UNIT FOR CHARGES

The practical unit of charge is the *coulomb*. It is the negative charge made by 6.25×10^{18} electrons. The term 10^{18} means it takes 6,250,000,000,000,000,000 electrons to produce a coulomb. Expressing it any other way than "6.25 times ten to the 18th" is awkward. (Some textbooks use 6.28 instead of 6.25. This is because 6.28 is 2π rounded off.)

THE VOLT

The volt is the unit for potential difference. It is used to indicate the electrical pressure or force needed to move coulombs of electric charge. The volt is also used to measure a unit of electromotive force (emf). The emf is the moving force behind an electric current. The volt is used and understood everywhere. The term *voltage* is often used to refer to potential difference.

CONTROLLING ELECTRONS

The controlled movement of electrons through a substance is called *current*. Current occurs only when a difference of potential is present. A good example of a difference of potential is observed by connecting battery terminals to a length of copper wire. The pressure from the battery moves the electrons.

Copper is a good path for current because of the relative ease with which electrons can be moved along its length. The one electron in the outer shell of the copper atom is free to move from atom to atom (see Fig. 1-8B). In fact, the electrons of copper drift in random fashion through the copper at room temperature (see Fig. 1-14A). If an imaginary line is set up in a copper wire, it will be found that the same number of electrons cross the line from both directions. This random movement does not produce an electric current. It takes a *controlled* movement of electrons to produce an electric current.

DIFFERENCE OF POTENTIAL (VOLTAGE)

Electric current results when the movement of electrons is in one direction (see Fig. 1-14B). This is done by applying a difference of potential or voltage across the ends of the wire. One end of the wire attracts electrons because it is connected to the

16 INTRODUCTION TO ELECTRICITY

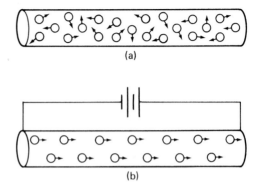

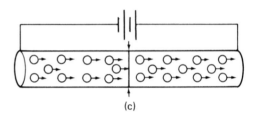

FIGURE 1-14 Current and the controlled motion of electrons. (A) Drifting electrons with no voltage applied. (B) The applied voltage controls the direction of electron flow. (C) The number of electrons past the line determines current flow.

battery terminal that has a positive charge or lack of electrons. The electrons in the copper wire drift toward this positive charge. As electrons leave the copper wire and enter the positive terminal, more electrons enter the other end of the copper wire. These electrons are taken from the negative terminal of the battery.

The difference of potential between the terminals of the battery is produced by a chemical reaction. When the chemical activity in the battery stops, the current stops.

ELECTRON FLOW (CURRENT)

Current is the rate at which electrons move. If a point is established in the copper wire (see Fig. 1-14C), the current can be measured by the number of electrons that pass this point each second. Recall that a certain number of electrons is a coulomb. When a coulomb of electrons moves past the spot in 1 second, this amount of current is 1 ampere. One *ampere* represents 6.25×10^{18} electrons passing a given point in 1 second. The current is 3 amperes when 18.75×10^{18} electrons pass a given point in 1 second.

CONDUCTORS

A conductor is a material that allows electrons to move easily. Copper is a good conductor because it has an electron far away from the nucleus that can be easily forced out of orbit. When the electrons in a material cannot be moved as easily as in copper, the material is said to present a higher resistance to the motion of charges. Good conductors are said to have a low resistance; poor conductors (called insulators) have a high resistance. When a voltage is applied to a material of high resistance (an insulator), there will be fewer electrons in motion and less current than if the same voltage were applied to a material of low resistance.

RESISTANCE

The ease with which electrons move in a material determines its resistance. A good conductor, such as copper, aluminum, or silver, has electrons that move freely. A low voltage will move a lot of electrons. A good insulator, such as glass, mica, or plastic, has electrons that do not move freely. Even a high voltage will move only a few electrons.

Resistance can have a wide range. It can be as low as that of a good conductor or as high as some good insulators. However, most resistances are somewhere in between good conductors and good insulators. The unit of measurement for resistance is the *ohm* (Ω). The *ohm* is defined as:

> One volt of pressure will push 1 coulomb of electrons through 1 ohm of resistance in 1 second.

Another way of saying it is that it takes 1 volt to push 1 ampere of electrons through 1 ohm of resistance.

THE ELECTRIC CIRCUIT

The workhorse of electricity is the circuit. It takes the electrons to where they belong or are needed. A complete circuit has a source of emf, a conducting path between the terminals of the power source, and a resistance, usually called the load. Note that all three elements, voltage, current, and resistance, are present in any complete circuit. And each has to be dealt with according to its presence.

The *series* circuit shown in Fig. 1-15 uses a battery, copper wire, and a light bulb. The battery produces the force needed to move the electrons. Chemical action in the battery makes the electrons available at the negative terminal. The copper wire is the path for the electrons to move along from the battery to the bulb. The copper wire is used because of its low resistance. Its resistance is less than 1 ohm.

It is necessary to have a complete path from one terminal of the battery to the other for electrons to flow. Electrons move only when there is a complete path between the two terminals. The second wire completes the path from the other end of the bulb to the positive terminal of the battery. This permits the electrons to return to the battery. The schematic for this circuit is shown in Fig. 1-16. This can be called a closed or complete circuit.

18 INTRODUCTION TO ELECTRICITY

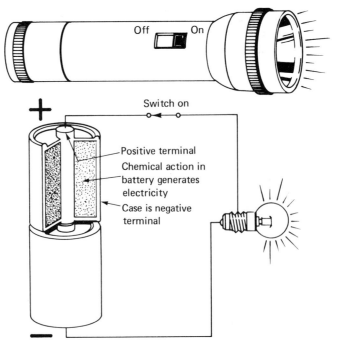

FIGURE 1-15 A simple series circuit has a battery and a bulb connected by two lengths of copper wire.

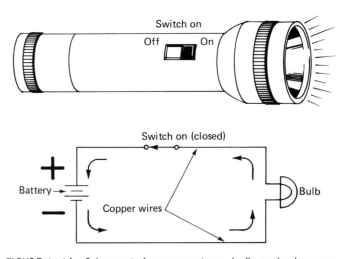

FIGURE 1-16 Schematic for connecting a bulb and a battery.

CONNECTING A CIRCUIT

When a circuit is connected to a battery, there is a negative charge at the negative terminal of the battery. Thus, an excess of electrons is located at this point. There is a positive charge at the positive terminal of the battery. Thus, there is a deficiency of electrons at this point. Electrons flow out of the negative terminal into the copper wire. This causes a movement of the electrons in one direction. Electrons also move in one direction in the bulb. The copper wire, connected to the other end of the bulb, conducts the electrons to the positive terminal of the battery. The arrival of the electrons at the positive terminal should end the movement of electrons. However, the chemical action in the battery maintains an emf across the battery terminals, and electrons continue to flow.

The light bulb has a resistance high enough to convert the electrical energy to light and heat energy. The electric current heats the filament of the bulb. It glows brightly. Keep in mind that light bulbs do not change all of the electrical energy into light. As you know, the light bulb also becomes hot. This means that much of the electrical energy is changed to heat.

Chemical action in the battery supplies the electrical energy to the light bulb. The chemical action in the battery has a limited lifetime. Eventually, the chemical action stops because the material in the battery is used up. When this happens, the light goes out because the battery is discharged.

SWITCHES CONTROL ELECTRON FLOW

Electron flow can be stopped by opening the circuit at any point. One of the wires can be removed from the battery terminal or removed from the bulb. In most electrical circuits, a switch is used at some point to permit the electron movement to be started and stopped when needed. Figure 1-17 shows a switch placed in series with the battery and the bulb. The switch can be placed at any point in the circuit.

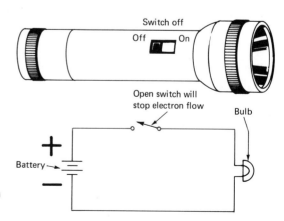

FIGURE 1-17 A switch is added in series with the bulb and battery to open and close the circuit.

SCHEMATIC

The *schematic* of Fig. 1-17 shows the switch in the open position. A schematic is a shorthand way of drawing a circuit using symbols. This switch can open and close the circuit without moving wires. This simple circuit diagram shows how a flashlight is wired.

REVIEW QUESTIONS

1. What is matter?
2. When two or more substances are mixed, what is the result called?
3. What is the result of chemically combining two or more elements?
4. Do all substances contain elements?
5. Which electrons in an atom affect its electrical characteristics?
6. All electrons are alike. (True or False)
7. The atom is made up of a nucleus and one or more _____.
8. Define electric current in terms of electrons.
9. What kind of charge does an electron have?
10. What happens to unlike charges?
11. What is the name given to an atom with a negative or a positive charge?
12. What type of electrons make up an electric current?
13. Define the coulomb.
14. Define the unit of measurement of electrical pressure.
15. What is another name for potential difference?
16. What kind of resistance does a conductor have?
17. One coulomb per second indicates what unit of resistance?
18. The unit of measurement for resistance is the _____.

PERFORMANCE OBJECTIVES

Understand the five ways electricity is produced.
Understand how the units of measurement for electricity were developed and used.
Be able to work Ohm's law problems.
Understand how volts, ohms, and amperes are related and function in any circuit.
Understand how to work electrical power problems.

CHAPTER 2

Current, Voltage, Resistance, Power, and Ohm's Law

SOURCES OF ELECTRICITY

The five most important sources of electricity for technicians are chemical action, heat, light, pressure, and magnetism.

Chemical Action

In the electrical and electronics fields, many sources of electricity are used. In the circuit of Fig. 2-1, the battery is the source of electricity. Batteries produce electrical energy by a chemical action.

Heat

Heat can be used to free electrons from some metals and from specially prepared surfaces. When some materials are heated to a high temperature, electrons are freed from their surfaces. Any nearby metallic surface, if positively charged, attracts these electrons and produces electron flow. The freeing of electrons by heat is called *thermal emission*.

Light

Light striking the surface of certain materials can be used to free electrons. This is called *photoemission*. With a suitable collecting surface, useful electron flow can result. Photoemission is used in photoelectric devices and television camera tubes.

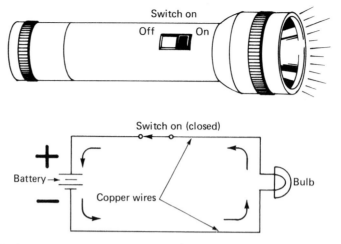

FIGURE 2-1 A battery produces current that flows in only one direction. Electrons move from negative to positive in the circuit, but positive to negative inside the battery.

Pressure

Mechanical pressure on certain crystals can be used to produce electricity. The crystal cartridge of an inexpensive record player is a good example. The needle causes a changing pressure. The crystal produces a changing voltage in step with the grooves in the record.

Magnetism

The most common method of generating electrical power is by turning a coil of wire in a magnetic field. This is the method the power station uses to generate the electric power that is used in homes, business, and industry.

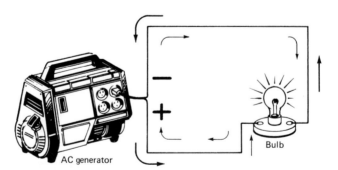

FIGURE 2-2 The output from an ac generator (alternator) produces a current that reverses direction at regular intervals. The current in the circuit alternates.

Electricity is generated in two forms: direct current (dc) and alternating current (ac). A battery supplies dc electricity because the electrons flow in only one direction, from the negative terminal to the positive terminal, as shown in Fig. 2-1. A battery supplies a constant voltage of one polarity.

The ac (alternating current) generator is shown in Fig. 2-2. It develops a voltage across its output terminals that changes polarity and amplitude as the coil is turned. At one time the electrons move in one direction (positive polarity). A fraction of a second later the electrons move in the other direction (negative polarity). These changes in current direction are called a hertz. The number of hertz completed in 1 second is called the frequency. In the United States, 60-hertz power is used. Some other countries use 50-hertz power. (Hertz is the newer term for cycles per second.)

When an electrical circuit such as the wires and bulb shown in Fig. 2-2 is connected across the output of the generator, the direction of electron movement depends on the polarity of the voltage. When the polarity is positive, the electrons flow from the negative terminal through the load to the positive terminal. A fraction of a second later the output voltage reverses. The electrons now flow in the direction shown by the heavy black arrows. The electrons in the circuit flow first in one direction and then in the other. Thus, the electron flow alternates. Since the current in the circuit changes direction, the voltage will also change direction. This is shown by the change in polarity (plus and minus) at the generator terminals.

There are advantages to using alternating current. One major advantage is that ac is easier to generate and distribute than dc. But many electronic circuits require both ac and dc. Such devices use power supplies to convert the ac electricity to the direct current used in the circuit.

UNITS OF MEASUREMENT

The importance of units for electrical measurements cannot be overemphasized. Electrical measurements are useful only if some standards of measurement exist. Let's review the electrical units.

Coulomb

The electrical charge on one electron has such a small value that it is not practical to measure it directly. A practical standard has been set up that says the unit of charge will be the total of 6.25×10^{18} electrons. This unit of charge is the *coulomb* (C). It is a basic measure of a quantity of electrical charge. Electron movement can also be measured in terms of coulombs if we include time as a second value. The electrical work performed by moving electrons depends on charge movement per unit of time. For example, a charge movement of so many coulombs per second is a rate of electron flow. This is a measurement of current or electricity at work. Remember, 1 coulomb per second is 1 ampere when it moves past a given point in 1 second.

26 CURRENT, VOLTAGE, RESISTANCE, POWER, AND OHM'S LAW

Ampere

A practical unit for measuring current is the *ampere*. It is used in place of the term *coulomb per second*. One ampere of electrical current is defined as the movement of 1 coulomb of electrons past a given point in 1 second. The ampere is measured with an ampere meter. This meter is called an ammeter.

Volt

The *force* that moves electrons is called a difference of potential, emf, or voltage. The unit of electrical potential is the *volt* (V). The volt is defined as the electrical force needed to produce 1 ampere of current in 1 ohm of resistance. The volt is measured with a voltage meter. This meter is called a voltmeter.

OHM

The opposition that a material presents to the flow of electrical charges is called resistance. The unit for electrical resistance is the *ohm* (Ω). By definition, the ohm represents the resistance a material offers to 1 ampere when 1 volt of electrical potential is present across the material.

The basic relation of units in an electrical circuit is shown in Fig. 2–3. If there are 2 volts present across the 2-ohm resistance, there will be a current of 1 ampere.

If the resistance remains the same and the voltage across the resistance is increased, the current will rise. Figure 2–4 shows the effect of increasing the voltage across the 2-ohm resistance. If the voltage stays at 2 volts, but resistance to the flow of electrons is increased by changing the resistance to 4 ohms, there will be a decrease in current. Figure 2–5 shows that the ammeter will indicate 1.0 ampere.

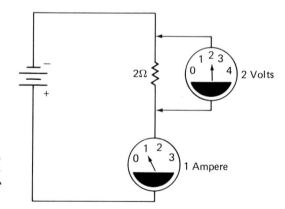

FIGURE 2–3 When 2 volts is present across 2 ohms, there is a current flow of 1 ampere through the resistor.

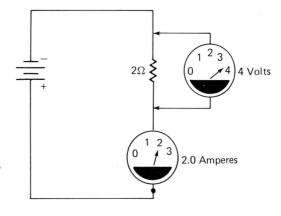

FIGURE 2-4 Changing the voltage to 4 volts across the 2-ohm resistor results in 2 amperes of current.

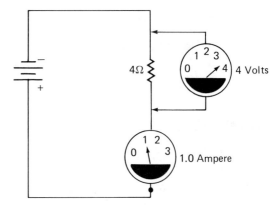

FIGURE 2-5 If 2 volts is present across 4 ohms, the current is 0.5 ampere.

PREFIXES

The ohm, volt, and ampere are units intended for practical use. For many uses, these basic units are either too large or too small. In such cases a system of prefixes is commonly used to make it easy to work with currents, voltages, and resistances.

Several of the more common prefixes are given in Tables 2-1 and 2-2. For example, a resistance of 10 million ohms can be written as 10,000,000 (10 × 1,000,000), 10 megohms, or 10×10^6 ohms. A current of 0.000003 ampere can be

TABLE 2-1
Common Prefixes and the Values They Represent

Prefix	Abbreviation	Multiply by:	Tens Power	Value
Mega	M	1,000,000.	10^6	Million
Kilo	k	1,000.	10^3	Thousand
Milli	m	0.001	10^{-3}	Thousandth
Micro	μ	0.000001	10^{-6}	Millionth

TABLE 2–2
Prefixes Used with Volts, Amperes, and Ohms

Prefix	Volts (V)		Amperes (A)		Ohms (Ω)	
M	1 megavolts	(1 MV)			10 megohms	(10 MΩ)
k	5 kilovolts	(5 kV)			5000 ohms	(5 kΩ)
m	10 millivolts	(10 mV)	5 milliamperes	(5 mA)		
μ	4 microvolts	(4 μV)	3 microamperes	(3 μA)		

represented as 3 microamperes or 3×10^{-6} ampere. A difference of potential of 5000 volts can be represented as 5 kilovolts or 5×10^3 volts.

These prefixes are not limited to the ohm, volt, or ampere. They are used frequently with other electrical units or values.

SIEMENS

The term conductance is also used often. It is the opposite of resistance. It is a term used to explain how well a wire or other substance conducts electricity. The unit of conductance is the *siemens* (S). The mho (ohm spelled backward) is the older term. Both are used in the literature, so it is best to be aware of both terms.

OHM'S LAW

Electric current is a flow of electrons. Electrons flow when a voltage is connected to a conducting path. The amount of current depends on the amount of voltage and the value of the resistance. Georg Ohm discovered the relationship between the three factors of voltage, current, and resistance. The exact values can be determined mathematically using Ohm's law.

Current is the result of an applied electrical force. The greater the applied electromotive force or voltage, the greater will be the current or amperes in a given circuit. Thus the voltage and current are directly related. In other words, *an increase in voltage causes an increase in current. A decrease in voltage will cause a decrease in current.*

The amount of resistance in the circuit also determines the amount of current that will flow. The lower the resistance (measured in ohms) in the circuit, the higher is the current (measured in amperes). Thus, current and resistance are *inversely* related. In other words, if one goes up the other goes down, and vice versa. Once you understand this concept, you also have a basic knowledge of Ohm's law. Ohm's law states that the current is equal to the voltage divided by the resistance.

$$\text{amperes} = \frac{\text{volts}}{\text{ohms}}$$

Usually, symbols are used for the units to form the equation:

$$I = \frac{E}{R}$$

where
- I = current in amperes (A)
- E = voltages in volts (V)
- R = resistance in ohms (Ω)

Keep in mind as you work Ohm's law problems that the basic units are the ampere, volt, and ohm. If you run into microamperes or milliamperes, they have to be converted to their decimal equivalent of the ampere. If you run into kilovolts, that unit has to be converted to volts. The same is true with resistance, which may be stated as *kil*ohms or *meg*ohms. You have to convert them to ohms before using the formula to obtain the correct answer.

Ohm's Law Examples

1. What is the current in amperes when the voltage applied is 100 volts and the resistance in the circuit is 50 ohms?

$$I = \frac{E}{R} = \frac{100}{50} = 2 \text{ amperes}$$

Table 2–3 shows the results of changes in voltage and current in a circuit having a fixed resistance of 50 ohms. Our answer, 2 amperes, is circled in the second column of Table 2–3, beside the 100-volt value used in the example.

If the voltage is doubled (increased from 100 to 200 volts), and the resistance remains fixed at 50 ohms, what happens to the circuit current?

$$I = \frac{E}{R} = \frac{200}{50} = 4 \text{ amperes}$$

TABLE 2–3 Relationship between Voltage and Current When the Resistance Is Held at 50 Ω and the Voltage Is Changed

Volts E	Amperes I	Resistance (R) = 50 Ω	
		Voltage Change	Current Change
400	8	4×	4×
350	7	3.5×	3.5×
300	6	3×	3×
100	②		
75	1.5	0.75×	0.75×
50	1	0.5×	0.5×
25	0.5	0.25×	0.25×
$I = \frac{E}{R}$			

When the voltage is doubled, or increased from 100 volts to 200 volts, and the resistance remains fixed at 50 ohms, the current is doubled. This is shown in Table 2-3 in the columns headed Voltage Change and Current Change. On the other hand, when the voltage is halved or decreased from 100 volts to 50 volts, what happens to the current?

$$I = \frac{E}{R} = \frac{50}{50} = 1 \text{ ampere}$$

When the applied voltage is decreased to half the original value, the current also decreases to half its original 2 amperes.

The *direct* relationship between voltage and current has been shown in this example. For a fixed amount of resistance, when the voltage goes up, the current goes up. When the voltage goes down, the current goes down. Thus, the current *varies directly* in step with the voltage.

The next step in understanding Ohm's law is to vary the resistance and keep the voltage constant.

2. What is the current when the resistance is changed to 75 ohms and the applied voltage is held at 100 volts? See Table 2-4.

$$I = \frac{E}{R} = \frac{100}{75} = 1.3333 \text{ amperes}$$

The resistance in the third column of Table 2-4 is 1.5 times the original value. The current in the fourth column is two-thirds of the original value. These values are circled in Table 2-4.

What is the current value for each of the resistances shown in the first column of Table 2-4? Fill in the fourth column (*I*) of the chart. Several answers have been filled in as examples.

Complete the third column of the chart to show the amount the resistance is increased or decreased. Fill in the fourth column to show the amount that the cur-

TABLE 2-4
Relationship between Resistance and Current When the Voltage Is Held at 100 V and the Resistance Is Changed

	Voltage (E) = 100 V		
Ohms R	Amperes I	Resistance Change	Current Change
150	1.500	3 ×	
100	1.000	2 ×	
(75)	(1.333)	(1.5 ×)	(0.666 ×)
50	2.000		
25	4.000	0.5 ×	2 ×
16.666	6.000	0.333 ×	3 ×
12.500	8.000	0.250 ×	4 ×
$I = \dfrac{E}{R}$			

rent changes. Note that the current decreases by the same ratio that the resistance increases, and the current increases by the same ratio that the resistance decreases. In these simple problems, you have shown that the current varies inversely with the resistance.

Ohm's Law and the Closed Circuit

It is important to have a grasp of the relation between current, voltage, and resistance. Ohm's law is the key to understanding these relationships. Ohm's law is a tool that can be used to determine what will happen in a circuit before it is turned on. A simple circuit is shown in Fig. 2-6. What happens in the circuit if a battery with a higher voltage is used? The current increases.

Assume that you have connected the circuit, measured the current, and have found it to be low. What do you know about the circuit? Low current means that the circuit resistance is high or the battery voltage is low. You know this because you know current is affected by voltage and resistance.

Suppose you want a certain current in a circuit (such as Fig. 2-6). What can

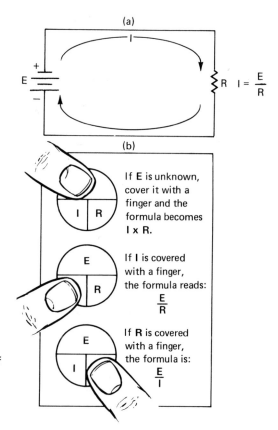

FIGURE 2-6 (A) The use of Ohm's law to plan what happens in a circuit.
(B) Using your finger to find the right formula to use.

32 CURRENT, VOLTAGE, RESISTANCE, POWER, AND OHM'S LAW

be done to produce this amount of current? You can change to a battery that has a different voltage. Keep in mind that a different voltage will produce a different current. Or you can change the resistance because a change in resistance also produces a change in current.

3. Determine the amount of current in a circuit when the resistance is 25 ohms and the voltage is 50 volts. Use Ohm's law to solve the problem:

$$I = \frac{E}{R} = \frac{50}{25} = 2 \text{ amperes}$$

The answer is 2 amperes. It has been entered in the Table of Fig. 2-7. Complete the table of values using 50 volts for each calculation. Place the values of voltage and current from the table in Table 2-5 on the graph of Fig. 2-7.

The point for two amperes at 25 ohms is shown on the graph. To place a point, find the value for 25 ohms across the bottom of the graph. The value 25 is marked by an arrow. Find the value for 2 amperes on the left side of the graph. Moving up from the 25 and across from the 2 will locate the point shown on the graph. Locate a point for each of the ohm and ampere values in Table 2-5.

The points you have plotted are now connected. Mark the line 50 volts. The line can now be used to find current and resistance values for a voltage of 50 volts.

If you should need to know the current when 50 volts is applied to a resistance of 12.5 ohms, you can find it by using this line. If you divide 50 by 12.5, you get 4 amperes. That is using Ohm's law for finding the value: $I = E/R$. The graph gives you the same answer. Electronics uses graphs for many functions. Get used to graphs for you will be seeing many of them.

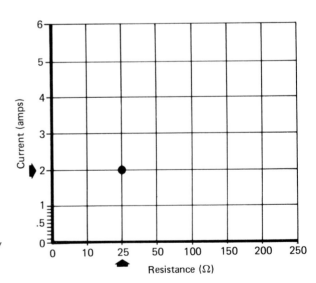

FIGURE 2-7 Graph to show the relationship of current, voltage, and resistance.

TABLE 2-5
Fill in the Values for Amperes Using Ohm's Law

$E = 50\ V$	$I = \dfrac{E}{R}$
R (ohms)	I (amperes)
10	
25	2.000
50	1.000
100	
150	
200	
250	

Ohm's Law in Other Forms

It is not always necessary to find the current in a circuit. If an ammeter is handy, it will give you this information. To be able to find the resistance or voltage in a circuit when one or the other is missing, it is possible to use Ohm's law to obtain the missing value. Keep in mind that $R = E/I$. And $E = I \times R$. These are nothing more than the other forms of Ohm's law. They come in handy when working with electricity.

If you want to find resistance when voltage and current are given, you use the formula $R = E/I$. For instance, if you have a 120-volt circuit that has 4 amperes flowing in it, the resistance can be found by dividing the voltage (120) by the current (4) to give the result of 30. This 30 is 30 ohms of resistance needed to cause the circuit to draw 4 amperes. Check Table 2-6 and fill in the blanks.

The relationship between voltage and current is evident when you use the resistance of 30 ohms and a higher current is found in the circuit. Note when you use

TABLE 2-6
Chart Showing the Relationship between Voltage and Resistance for a Current of Four Amperes

$I = 4\ amperes$	
E	R (Ω)
600	
480	
360	90
240	
120	30
60	
40	
30	7.5
24	6.0
$R = \dfrac{E}{I}$	

TABLE 2-7 Chart of Current and Voltage for a Constant Resistance of 30 Ohms

$R = 30\,\Omega$,	$E = I \times R$
I	E
16	
12	
8	
4	120
2	
1.333	
1	30

$E = I \times R$ that the voltage is directly related to the current. If the current is increased and the resistance stays fixed, the voltage has to go up or increase.

If you have 4 amperes of current and 30 ohms of resistance you use the formula: $E = I \times R$ and get $E = 4 \times 30$ or 120 volts. Place this value in Table 2-7. Now, see if you can fill in the second column of the chart in Table 2-7.

Uses for Ohm's Law

It is sometimes awkward to express values in basic units because many zeros or decimals are needed. For this reason, several prefixes and symbols are used. These are shown in Table 2-8. These symbols are often used in making Ohm's law calculations. They are also used for electronic terms such as frequency, inductance, and capacitance, which you will encounter in later chapters.

Most circuits have the voltage stated in volts. High-voltage cross-country transmission lines and television picture tube voltages are usually given in kilovolts or thousands of volts. Kilo (k) means 1000. Thus, 138,000 volts can be written as 138 kilovolts and abbreviated to 138 kV.

TABLE 2-8 Prefixes and Multipliers for E, I, and R

Prefix	Symbol	Multiple to Get Basic Unit	Multiplication Factor	E	I	R
tera	T	1,000,000,000,000	10^{12}			
giga	G	1,000,000,000	10^{9}			
mega	M	1,000,000	10^{6}	megavolts		megohms
kilo	k, K	1,000	10^{3}	kilovolts	kiloamperes	kilohms
hecto	h	100	10^{2}			
Basic Units		1	10^{0}	Volt	Ampere	Ohm
centi	c	0.01	10^{-2}			
milli	m	0.001	10^{-3}	millivolts	milliamperes	
micro	μ	0.000001	10^{-6}	microvolts	microamperes	
nano	n	0.000000001	10^{-9}	nanovolts	nanoamperes	
pico	p	0.000000000001	10^{-12}			
femto	f	0.000000000000001	10^{-15}			
atto	a	0.000000000000000001	10^{-18}			

POWER

Power can be understood when you take the knowledge you have and use it with Ohm's law formulas and examine them in terms of the work done. The formulas are modified slightly to produce the answers needed when figuring the work done by electricity in a circuit.

Work is done when you lift a 100-pound object for a distance of 10 feet. Power is the rate of doing work, or *foot-pounds per second.* If you take 20 seconds to lift the 100 pounds 10 feet, then you can find the amount of work done in *foot-pounds.* Power is equal to foot-pounds divided by time. In this case you have lifted 100 pounds for a distance of 10 feet. Thus, you multiply the 100 by 10 to produce 1000. This 1000 is divided by the time it took to do the job (20 seconds) to produce 50 foot-pounds of work.

If you want to convert foot-pounds to horsepower, a term you are more familiar with, you keep in mind that 1 horsepower is equivalent to 550 foot-pounds per second, or lifting 550 pounds for 1 foot in 1 second.

The 50 foot-pounds is about 0.0909 horsepower, or it can be said in fractional form as 1/11th of a horsepower.

JOULE

Electrical work (joule) is done when force (E) causes the movement of an electrical charge (Q). This can be stated as joule = $E \times Q$.

The unit of electrical work is the *joule.* One joule represents the work done by a difference of potential of 1 volt (E) while moving 1 coulomb (Q) of charge (6.25×10^{18} electrons).

Power (W for watts) in an electrical circuit is the rate of doing electrical work (volt-coulombs per second).

As an equation, it becomes

$$power = \frac{volts \times coulombs}{time}$$

or

$$P = \frac{EQ}{t}$$

where

P = charge
E = voltage
Q = power
t = time

WATT

The unit of *electrical work* is the *watt* (W). The letter *P* is used to represent power. One watt is equal to 1 joule per second. Thus, 1 joule and 1 watt are the same and you can convert back and forth between joules and watts easily. It takes 746 watts of electrical energy to produce 1 horsepower.

$$1 \text{ hp} = 746 \text{ watts}$$

PUTTING ELECTRICITY TO WORK

Many jobs can be done by electricity. One useful conversion is from electricity to light. We depend on lights for the home, automobile, street lights, and flashlights. The most common conversion from electrical energy is to heat. An electric toaster or an electric iron such as in Fig. 2-8 provide useful forms of heat energy. Heat is produced in most electrical conversion devices. In some, such as the light bulb, heat is an unwanted result.

RESISTORS AND HEAT

Resistors, such as those shown in Fig. 2-9, are used in electric circuits. They act as loads, voltage dividers, and current-control elements. However, in doing their job, heat is produced. For this reason, electrical circuits must be ventilated to carry away

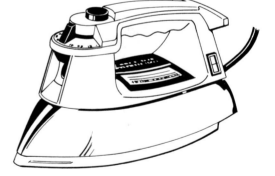

FIGURE 2-8 The electric iron is an example of a device that converts electrical energy into heat.

FIGURE 2-9 A resistor placed across a voltage source gets hot due to the current in the resistor.

the heat. Also, a resistor must be able to handle the power so that it does not overheat and burn.

MECHANICAL ENERGY

Electrical energy is often converted to mechanical energy. The electric motor shown in Fig. 2-10 is a good example of a device used to convert electrical to mechanical energy. Another example is a relay, which is a switch that is operated electrically. When electrical energy is applied to the relay, a movable metal arm turns on or off another electric circuit. A relay uses a low-voltage, low-current source to control high voltage or high current from a remote location. Furnaces, air conditioners, and refrigerators all use relays.

Many electrical devices are rated in terms of power and operating voltage. It is often necessary to find the current that results when the device is connected to a certain voltage source. At other times the power and the current are known, and it may be helpful to find how much voltage is needed to operate the device. Both of these problems can be solved by using one of the forms of the basic power law:

$$P = E \times I$$

$$P = I^2 R$$

$$P = E^2/R \quad \text{or} \quad P = \frac{E^2}{R}$$

You can also convert these formulas to

$$I = P/E \quad \text{or} \quad I = \frac{P}{E}$$

$$E = P/I \quad \text{or} \quad E = \frac{P}{I}$$

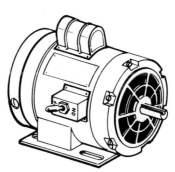

FIGURE 2-10 A motor is often used to convert electrical energy to mechanical motion.

Or you can use $I^2 = P/R$ and convert it to

$$I = \sqrt{P/R} \quad \text{or} \quad I = \sqrt{\frac{P}{R}}$$

A good example of how this formula can be used is found in the following problem: An 800-watt electric heater has a resistance of 2 ohms. What is the current needed to operate the heater?

First, select the formula needed. This is determined by the known values. In this case, the power and the resistance are known. You need to find the current. So, you select $I = \sqrt{P/R}$ for a formula, substitute the values in the formula, and solve.

$$I = \sqrt{P/R} \quad \text{or} \quad \sqrt{800/2} = \sqrt{400} = 20 \text{ amperes}$$

This is illustrated in Fig. 2-11.

Knowing that the required current is 20 amperes, you can determine the size of the wire needed to cause the heater to operate properly. You look up the current in a chart and find that No. 12 copper is the wire size for safely handling this amount of current.

KILOWATT-HOUR

Public utilities that furnish power for your use charge for electrical energy by using a popular unit of electric work, the kilowatt-hour (kWh). One kilowatt-hour means you used 1000 watts of electric power for 1 hour. If the cost of power is 10 cents per kilowatt-hour, and if you operate a 1000-watt iron for 1 hour, you will be billed for 10 cents. If you operate the iron for 6 hours, you will use 60 cents worth of electrical energy.

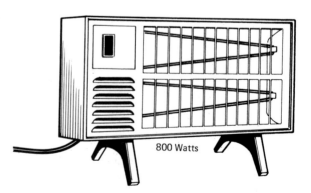

FIGURE 2-11 An electric heater has low resistance and high current.

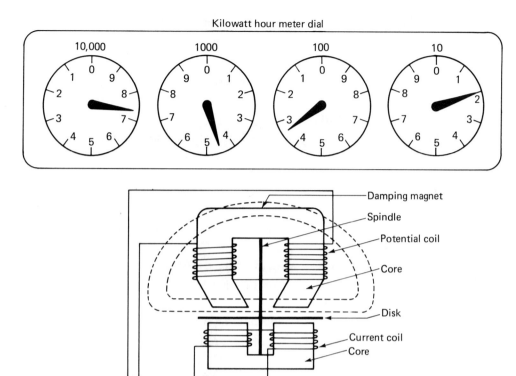

FIGURE 2-12 Kilowatt-hour meter. Note how it is made and how the dials are read from left to right.

The kilowatt-hour meter is a motor-type device that accumulates and displays the kilowatt-hour usage, which is then read by a meter reader and subtracted from the last reading to see how much was used in the interval between meter readings (see Fig. 2-12).

One *kilowatt* is 1000 watts. One kilowatt-*hour* is 1000 watts for 1 hour. If the kilowatt-hour is not specified, it means the power being consumed in one *second*.

REVIEW QUESTIONS

1. The unit of measurement for conductance is _____.
2. Electricity from a battery is produced by a _____ action.
3. How will pressure on crystals produce electricity?

4. What kind of current is produced by using magnetism?
5. Does a battery produce alternating or direct current?
6. The ammeter is used to measure _____.
7. The resistance of a circuit is decreased; what happens to the current if the voltage remains the same?
8. What is a complete circuit?
9. How much current will flow if 1 coulomb per second passes a given point in 1 second?
10. Electrical pressure is measured by a _____ meter.
11. What happens to the current in a circuit when the emf applied to a resistance is increased?
12. What happens to the current if the resistance is decreased and the voltage is not changed?
13. Determine the current when the voltage is 64 volts and the resistance is 32 ohms.
14. A voltage applied to a resistance is 64 volts. Resistance is increased from 32 ohms to 64 ohms. What happens to the current?
15. A voltage of 60 volts is applied to a resistance of 20 ohms. What is the current?
16. A voltage of 45 volts is applied to a resistance of 10 ohms. What is the current through the resistor?
17. The current in a resistance is 2 amperes. It is across a 12-volt battery. What is the value of the resistance?
18. A current of 3 amperes is present in a 50-ohm resistance. What is the applied voltage?
19. What is the voltage drop across a resistor that has 8 ohms of resistance and 1.5 amperes of current through it?
20. What is a siemans?
21. How is a siemans related to an ohm?
22. What is the unit of measurement for electrical work?
23. What is the unit of measurement for power?
24. There is a movement of 50 coulombs in 20 seconds; what is the power if the voltage is 6 volts?
25. What is the current if 50 coulombs is moved in 20 seconds by 6 volts of electrical pressure?
26. What is the basic power formula?
27. What is the rate of doing work called?
28. The current in a 50-ohm resistance is 5 amperes. Find the voltage drop and power used.

29. The current in a 30-ohm load is 2 amperes. Find the voltage drop and power used.
30. A 300-watt bulb burns for 4 hours. What is the power consumed in kilowatt-hours?
31. What is the formula used to find power if the voltage and resistance are known?
32. What is the formula for finding power if the current and resistance are known?

PERFORMANCE OBJECTIVES

Be able to read the resistor color code.

Be able to recognize variable and fixed resistors.

Understand how the capacitor is used in air-conditioning and refrigeration circuits.

Know the difference between a standard capacitor and an electrolytic capacitor.

Know the difference between an inductor and a transformer.

Know that semiconductors such as diodes, transistors, and silicon chips are used in electronic controls for heating, air-conditioning, and refrigeration devices.

CHAPTER 3

Resistors, Color Code, Components, and Symbols

RESISTORS

Resistance is found in every electrical circuit. Voltage is the pressure that pushes the electrons through the circuit and through the resistance of the consuming device. Everything has resistance to some degree. It is possible to use certain substances in various configurations to produce a device that will limit the amount of current in a circuit. This device is called a *resistor*. It becomes very useful when you want a number of pieces of equipment to operate from one voltage source. It would be very difficult to use batteries of different voltages for every component in an electrical circuit. Resistors of various sizes can be placed in a circuit to cause it to have the correct voltage for the various devices. Resistors are used to drop voltage.

Resistors of various sizes are available. It takes a few pages of any electronics supply house catalog to list all the types handled by the supplier. Our purpose here is to take a look at the method used to mark the value on carbon-composition resistors.

Carbon-composition resistors are small in physical size. They can dissipate $\frac{1}{8}$, $\frac{1}{4}$, $\frac{1}{2}$, 1, and 2 watts. Thus, some method must be devised to mark them. If they are placed in a circuit, it is best to be able to read the value without having to turn the resistor over. Turning it over after it has been soldered in place can cause the connection to break or the leads to break off. Color banding of the resistor seems to be the best answer to the problem of marking.

Carbon-composition resistors have been color coded for a number of years. Three color bands are used to indicate the ohmic value of the resistor. A fourth band is used to indicate the tolerance (see Fig. 3-1).

44 RESISTORS, COLOR CODE, COMPONENTS, AND SYMBOLS

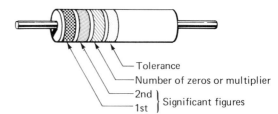

FIGURE 3-1 Color bands on a carbon composition resistor.

Resistance is measured in ohms. Wire-wound resistors normally have their values (in ohms) marked on them, since they are physically larger than the carbon-composition type and the lettering or printing can be made on the body of the resistor.

COLOR CODE

The colors used on carbon-composition resistors have a definite value, which is easily read. Each color has a value assigned to it, which is agreed upon internationally, as follows:

Color	Number
Black	0
Brown	1
Red	2
Orange	3
Yellow	4
Green	5
Blue	6
Violet (purple)	7
Gray	8
White	9

When looking at the resistor, you will note that the bands are located at one end. You start reading the resistor value by checking the band nearest the end of the resistor first. Turn it so that you are holding the end with the color band to your left. Then you can read from left to right, as you usually do.

The first two color bands, A and B in Fig. 3-2, indicate the first two digits in the resistance value. The third band is used to show how many zeros are added. C

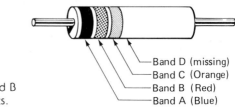

FIGURE 3-2 Bands A and B indicate the first two digits.

shows how many zeros follow the first two digits, or it is the multiplier. Sometimes the fourth band (D) is missing. If so, the resistor has a tolerance of ±20%.

Tolerance

When there is a fourth band, it will be either silver or gold. Silver means the tolerance of the resistor is ±10%, while gold indicates ±5%. The lower the tolerance number (5% instead of 10%) the more expensive the resistor is to produce and also buy. The tolerance tells how close the resistor value is to that indicated by the color bands.

Tolerance is indicated in + or − designations to tell you how accurate the resistor is. It may be in the + direction or have *more* resistance than the value stated by the color bands. Or it may have a − value, meaning the value of the resistor will measure *less than* the color bands indicate.

Examples

A few examples will make you more familiar with the code and its usefulness. Examine Fig. 3-3. Note the first three bands. The blue–red–orange bands signify 62 followed by three zeros. This produces a value of 62,000 ohms. There is no fourth band, so the tolerance is 20%.

What are the *limits* of the resistor when you use a meter to check its value? Take the 62,000 and multiply it by 0.20. That produces 12,400. You add the 12,400 to the 62,000 to get 74,400 ohms for the + tolerance value. You take 12,400 away from the 62,000 ohms to get the − or lowest value the resistor can have to still be within tolerance. So, 12,400 from 62,000 produces 49,600 ohms. Thus, the resistor can read between 49,600 ohms and 74,400 ohms and still be called a 62,000-ohm resistor as indicated by its color bands.

Let's take another look at the color code and what it can do for us. Check Fig. 3-4. In this figure you have three color bands to obtain the value of the resistor. The violet (7), green (5), and red (2) indicate that the resistor has a value of 7500 ohms. The gold band indicates it has a value of 7500 ±5%. Five percent of 7500 ohms is 375 ohms. That gives the 7500-ohm resistor a tolerance range of 7500 + 375 and 7500 − 375. The results are 7875 ohms and 7125 ohms. The resistor can read anywhere between 7125 and 7875 and still be called a 7500-ohm resistor.

Sometimes you may find a resistor such as shown in Fig. 3-5. The resistance

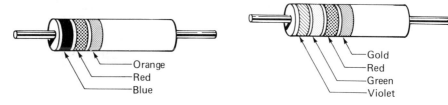

FIGURE 3-3 A 62,000-ohm resistor.   FIGURE 3-4 A 7500-ohm resistor.

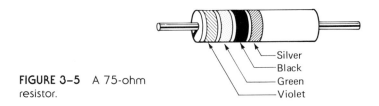

FIGURE 3–5 A 75-ohm resistor.

value is not 750 ohms, but 75 instead. The third band specifies the number of zeros. Since black is zero, there are *no zeros* after the first two digits. Thus, the black adds nothing and the 75 stands alone. The silver tolerance on the resistor indicates that it has a plus or minus value of 7.5 ohms (75 times 0.10 produces 7.5 ohms). The 7.5 ohms has to be added to and subtracted from 75 to produce the tolerance range of the resistor. So, 75 plus 7.5 equals 82.5 ohms, and 75 minus 7.5 equals 67.5 ohms. The resistor has a tolerance range of 67.5 to 82.5 ohms.

Gold and Silver Third Bands

Carbon-composition resistors have been improved to the point where they can be made in values of less than 10 ohms. Thus, the code has to be altered to fit the situation, and using gold or silver as the third band comes in handy.

Gold as the third band means you divide the first two numbers by 10. Thus, if you have a resistor of red, red, and gold, it has a value of 22 divided by 10, or 2.2 ohms. As you can see, this makes it possible to use the color code now for indicating resistances of less than 10 ohms.

Examine Fig. 3-6. Note that the values of the resistor are indicated by a first band of blue and a second band of yellow, which produces 67. The third band is gold. So, the 67 is divided by 10 to produce a reading of 6.7 ohms for the resistor. If the fourth band is silver, it means the tolerance is ±10%. Ten percent of 6.7 ohms is 0.067 ohm. Add the 0.067 to the 6.7 and you have 6.767 ohms. Subtract 0.67 ohm from the 6.7 and you have a negative tolerance value of 6.633 ohms. The tolerance range is 6.633 ohms to 6.767 ohms. It takes a very good ohmmeter to measure this close a tolerance. In fact, it would be hard to see anything near the 6.7 on a conventional ohmmeter. It would probably read 7 or even as much as 10 ohms. The digital ohmmeter is used to measure this type of ohmic value.

Silver as the third band means you divide the first two numbers by 100. Thus, if you have a resistor of red, red, and silver, it has a value of 22 divided by 100, or 0.22 ohm. Take a look at Fig. 3-7. The resistor has bands of yellow, violet, and

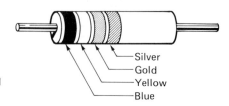

FIGURE 3–6 Note the third band is gold.

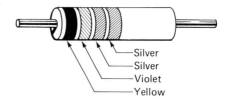

FIGURE 3-7 Note the third band is silver.

silver. The values then are 47 divided by 100 to produce the correct answer of 0.47 ohm for the resistor. Most of these resistors will have a fourth band, also. The fourth band still indicates the tolerance. So, if the resistor has a fourth band of silver, it is ±10% as usual. This means that the resistor will vary in resistance by 0.047 ohm in either direction of 0.47. Thus resistor tolerance may be from 0.423 to 0.517 ohm. It takes a very good ohmmeter to measure this resistance range; but the advent of semiconductors has brought about more sensitive and inexpensive meters and the demand for lower resistance values and closer tolerances.

Resistors come in standard sizes. The Electronics Industries Association (EIA) sets standards for the manufacture of resistors. This helps to standarize the number of sizes of resistors made. It also makes it easy to find spare parts to use for repair jobs.

VARIABLE RESISTORS

Some resistors are variable. This means that the amount of resistance can be changed. Variable resistors may be either carbon composition or wire wound.

These resistors are used for special circuits. On these circuits, the amount of voltage or current that is delivered must be varied. A common example is the volume control on your radio or television set (see Fig. 3-8).

Variable resistors are easily identified because they have three connections for leads. The center lead is usually the variable contact. A variable resistor that is connected into a circuit at all three points is called a *potentiometer* (see Fig. 3-8). A potentiometer is often referred to as a *pot*.

Usually, a potentiometer is used to vary voltage. The device is connected across a voltage source by placing it directly across the battery or power source. The variable arm is then used to change the voltage that is available from the potentiometer.

The *rheostat* is a variable resistor. It is used by connecting it in series not across

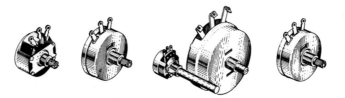

FIGURE 3-8 Potentiometer.

48 RESISTORS, COLOR CODE, COMPONENTS, AND SYMBOLS

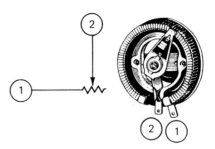

FIGURE 3-9 A rheostat symbol. Rheostats resemble potentiometers. A potentiometer can be made to serve as a rheostat by connecting the center terminal and one of the end terminals to form a single connection.

the voltage source, as was the case with the potentiometer. Rheostats are designed to handle higher currents than potentiometers. Very few rheostats are used today because their jobs are being done by semiconductors. Usually, a rheostat is connected to a circuit at only *two* points. A symbol for the rheostat is shown in Fig. 3-9.

Variable resistors have a wide range of adjustments. For example, volume controls typically use carbon resistors. Resistance ratings can be adjusted from 0 to 10 million ohms. Another way to state these values is from 0 to 10 megohms (mega means million).

Many potentiometers have what is called a *nonlinear* resistance element. This simply means that resistance does not change at a fixed, or uniform, rate as adjustments are made. Usually, they are small, or fine, changes at the low end. At the high end, settings lead to large resistance changes. This nonuniform resistance leads to what is called a tapered control. Such devices are usually used to adjust sound volumes and are called audio taper resistors.

There are also *linear taper* potentiometers. They have a uniform change of resistance as the settings are adjusted. They look exactly the same as the audio taper. When replacing a potentiometer, you must be very careful not to use a linear taper one in a volume control circuit or, worse yet, an audio taper in a control circuit. This is one of the things that you, as a technician, must be aware of in making repairs. Do not try to substitute a volume control of the same resistance for a control circuit potentiometer. You will find it very difficult to make the required adjustments in the control circuit.

Wattage ratings are usually marked on the rheostat or potentiometer. It is difficult to tell the wattage rating by just observing the device. It takes practice to be able to tell the difference between various wattage ratings.

SCHEMATIC

Persons who work with electricity use a *schematic* to tell them where components are located and what the relationships are to one another. A schematic is a diagram something like a road map. It tells you the way from one point to another. Symbols are used to mark certain items along the way. The schematic is the road map of the electrical trade. Many types of diagrams are needed to describe the operation and construction of electrical equipment. The schematic is the most common term for the electrical diagram.

The schematic shows how the electrical or electronic components (parts) are connected. A schematic is a handy item to have when you are analyzing, explaining, or servicing heating, air-conditioning, and refrigeration equipment.

TYPES OF RESISTORS

The resistor is the most widely used electrical and electronic device. Every radio, television set, and control circuit has a resistor or resistors. This component is used to provide resistance. It is designed to be used at a fixed value or as a variable-value device.

Fixed Resistors

The fixed resistor is the simplest of the two types. It is made so that you cannot change the resistance. Some carbon fixed resistors are shown in Fig. 3-10. These are carbon composition and have a cover of black, brown, or green plastic. A color code is used to give the value of the resistor.

Fixed wire-wound resistors are available for use when the wattage rating is higher than 2 watts. Carbon-composition resistors come in $\frac{1}{8}$, $\frac{1}{4}$, $\frac{1}{2}$, 1, and 2 watt sizes. The physical size tells the rating. You get used to the wattage rating when working with resistors. The larger the resistor, the higher the wattage rating is. The larger the resistor, the easier it is for it to dissipate heat. Since resistors put up resistance to current flow, they also drop voltage. The energy has to be dissipated as heat. Thus, the surface of the resistor must be large enough to allow the heat to be dissipated.

Figure 3-11 shows some fixed wire-wound resistors. These are made of high-

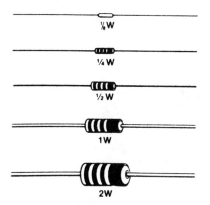

FIGURE 3-10 Fixed carbon-composition resistors.

FIGURE 3-11 Fixed wire-wound resistors.

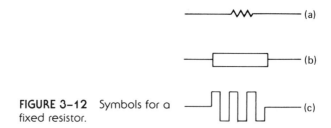

FIGURE 3-12 Symbols for a fixed resistor.

resistance wire wound on an insulating core with a ceramic coating. They usually are large enough so that the resistance of the unit can be stamped on it.

The symbol for a fixed resistor is shown in Fig. 3-12. Note how the symbols vary for different users. A is the standard EIA (Electronics Industries Association) symbol. B is usually used by foreign manufacturers and occasionally by American makers of industrial equipment. C is seldom encountered, but it is sometimes used in schematics for industrial equipment and can be seen in some refrigeration and air-conditioning electrical schematics.

Tapped Resistors

Tapped resistors are used in some circuits. They have taps for easy connections. They are usually wire wound, although some are carbon. Figure 3-13 shows samples of the tapped resistor. Figure 3-14 shows the schematic representation of tapped resistors.

The ceramic coating is left off the wire where the tap is to be made. This allows a sliding connection so that the tapped resistor can be made into a variable resistor or adjusted as needed.

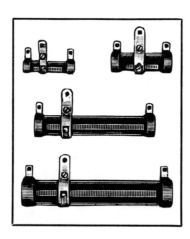

FIGURE 3-13 Tapped resistors.

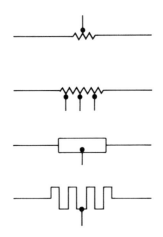

FIGURE 3-14 Schematic representation of tapped resistors.

Variable Resistors

A variable resistor may be made of carbon or it may be wire wound. The idea behind the variable is to make it adjustable to meet the needs of the circuit. You are most familiar with the variable resistor as a volume control on a radio or television set. This is a variable carbon-composition type of resistor and controls a circuit to allow for increases or decreases in volume, as you desire.

A variable resistor has a movable contact that is used to adjust or select the resistance value between two terminals. In most uses, the variable resistor is a control device. It is made in many sizes and shapes. Figure 3-15 shows some of these types. The shafts of most variable resistors have knobs placed on them to make them easier to use. However, some are made to be adjusted by the insertion of a screwdriver blade in a slot on the resistor. Many adjustable resistors are used in controls for air-conditioning and refrigeration systems. Figure 3-16 shows the schematic representation for variable resistors.

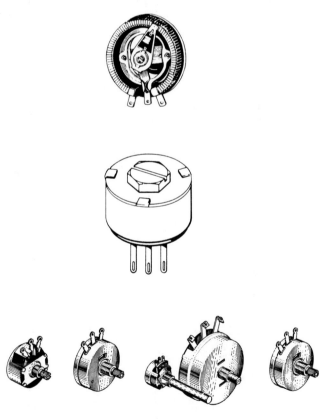

FIGURE 3-15 Types of variable resistors.

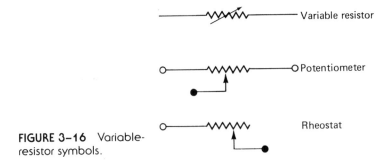

FIGURE 3-16 Variable-resistor symbols.

Fusible Resistors

In some cases, the resistor has a purpose other than providing resistance. One type is used to protect the equipment or circuit against excess current surges. This type of resistor, called a fusible resistor, is built to fail before damage is done to more expensive parts. Such units are often made to plug into a socket (see Fig. 3-17). Figure 3-18 shows the schematic symbol for a fusible resistor.

Temperature-compensating Resistor

Another type of special resistor is the temperature-compensating resistor. It is designed so that the resistance value changes in a direct or inverse relation with temperature changes. Such resistors are used to provide special control of circuits that must be extremely stable in operation. The symbol is shown in Figure 3-19.

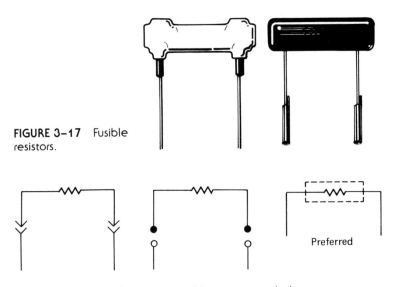

FIGURE 3-17 Fusible resistors.

FIGURE 3-18 Fusible-resistor symbol.

FIGURE 3-19 Symbol for temperature-compensating resistor.

TYPES OF CAPACITORS

The capacitor is used in many electrical circuits in both electronics and in air-conditioning and refrigeration circuits. Two types of capacitors are used in these circuits: fixed and variable.

Fixed Capacitors

The fixed capacitor is made for a certain value and is not adjustable. The fixed capacitor is divided into several groupings. It may be made with paper separating two plates of aluminum foil, or it may use plastic, mica, ceramic, or electrolytes.

Most paper capacitors have been replaced by those made of better materials, usually plastic. A typical capacitor is shown in Fig. 3-20. Capacitors are large

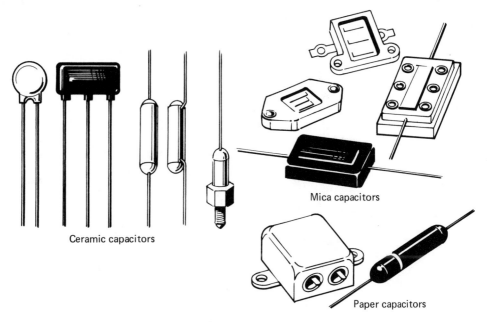

FIGURE 3-20 Fixed capacitors.

enough to have their values printed on them. The smaller capacitors use a color code to indicate their value and working voltage. Capacitors come in hundreds of sizes and shapes. It takes a good half-hour to thumb through an electronics catalog that shows all the various types. Each type has a special or particular application. Mica types, for instance, are used for some high-frequency applications with high voltages. The ceramic type is found in circuits that use high voltages, such as television sets and radar equipment.

Electrolytic Capacitor

This type of capacitor needs special mention, but will be covered in more detail in Chapter 10. The electrolytic capacitor has a very high capacitance value when compared with the types mentioned previously. These capacitors may be tubular or square in shape. They have cardboard or metal covers. Values are printed on the cardboard cover and stamped into the metal cover (see Fig. 3-21). They are available in a variety of shapes and sizes. One characteristic of the electrolytic capacitor is its polarity. Its terminals will have − (negative) or + (positive) marked on them. This means that the circuit power must be connected correctly to avoid damage to the electrolytic capacitor. It is not to be used on ac unless it is an ac electrolytic capacitor and so identified.

If an electrolytic capacitor marked with a − and a + is connected to ac, it will explode, and can throw its contents over an area as large as 50 square feet. Thus it can be dangerous. Some are manufactured with a small hole in them so that their contents will spew out instead of exploding. However, safety dictates that you treat all electrolytic capacitors as firecrackers and a larger one as a piece of dynamite. This is another reason for wearing eye protection when working with electrical circuits.

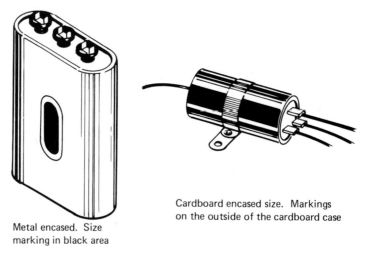

Metal encased. Size marking in black area

Cardboard encased size. Markings on the outside of the cardboard case

FIGURE 3-21 Electrolytic capacitor with value marked on it.

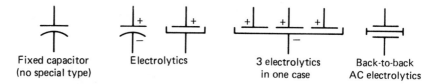

FIGURE 3-22 Capacitor symbols with electrical polarity marked.

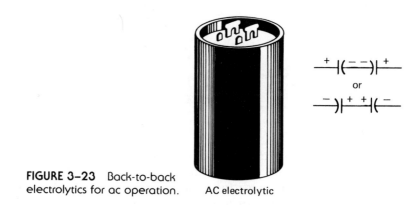

FIGURE 3-23 Back-to-back electrolytics for ac operation.

Note that the symbols for capacitors in Fig. 3-22 indicate the electrolytic capacitor with polarity markings.

Electrolytic capacitors are 1 microfarad (μF) and larger in size. They can be made to operate on ac by connecting two of them back-to-back as shown in Fig. 3-23. AC electrolytic capacitors are used in electrical motors, cross-over networks in speaker systems, and other places needing large capacitances in circuits that contain ac.

Variable Capacitors

Variable capacitors are used for tuning purposes in radios and televisions. In most instances, you will not need them for air-conditioning and refrigeration circuits. However, in case you do see one utilized in the electronics control unit, you can identify it by using Fig. 3-24.

TYPES OF INDUCTORS

Coils, inductors, and chokes are the names used to indicate a coil of wire. "Inductor" is preferred because inductors have inductance, a property that is utilized in many electrical circuits.

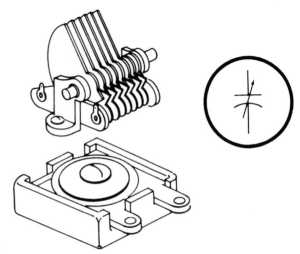

FIGURE 3-24 Variable capacitors with symbols.

Fixed Inductors

The simplest coil or inductor has an air core and is made by winding a wire in a series of loops, which may or may not have a form to hold them in place. Coils are seldom color coded for value, so we look at the schematic or a parts list for the inductance value of a coil. Inductance is the electrical property of a coil, just as resistance is the electrical property of a resistor. Many coils are wound on plastic forms that support the loops of wire. The form has no effect on the operation of the coil. The symbols for air-core coils are shown in Fig. 3-25. Other types are powdered iron core and iron core. Symbols for these types are shown in Fig. 3-26.

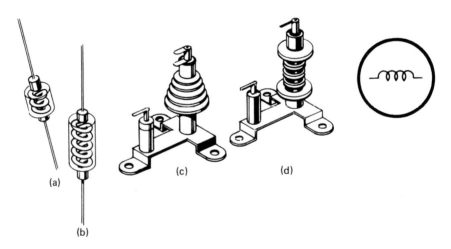

FIGURE 3-25 Symbols for air-core coils.

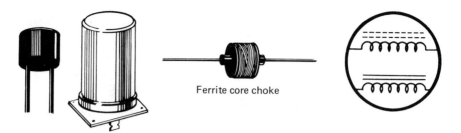

FIGURE 3-26 Powdered-iron-core and iron-core inductors.

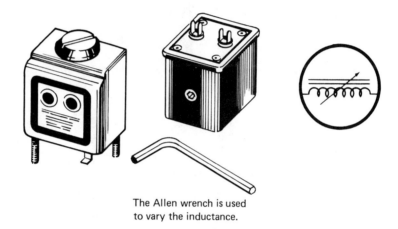

The Allen wrench is used to vary the inductance.

FIGURE 3-27 Symbol for variable inductors.

Variable Inductors

Some circuits need inductors that can have their values changed, some by screwdriver adjustment and others by changing the core material. Figure 3-27 shows the symbols for variable inductors. Note the differences for iron-core variable inductors. Iron-core (made of iron or steel) chokes are indicated by two straight lines over the loops. Dashed lines indicate powdered iron cores.

TRANSFORMERS

The inductor has one coil. The transformer has two. The two coils of a transformer are so placed that the energy in one is transferred to the other by magnetic induction.

The coil winding connected to the power source is called the *primary*. The winding connected to the circuit or consuming device is called the *secondary*. Or keep in mind that the primary is the input and the secondary is the output. Note

58 RESISTORS, COLOR CODE, COMPONENTS, AND SYMBOLS

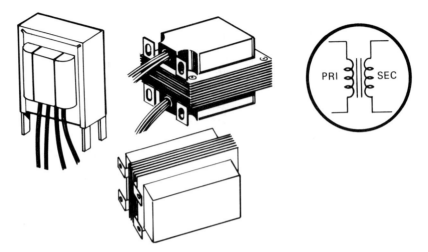

FIGURE 3-28 Symbol for a transformer.

in Fig. 3-28 that the two coils are not connected. Transformers are used in air-conditioning and refrigeration circuits for stepping down the voltage from 120 volts or 240 volts to 24 volts. The thermostat circuit then uses the 24 volts. The transformers you will work with have iron cores. Air-core transformers are used in electronics where radio frequencies are present.

SEMICONDUCTORS

You know semiconductors as transistors, diodes, and chips. Silicon material is used to make devices that will conduct a certain amount of current when needed. These devices are made for easy use and with low voltages. You will encounter semiconductor chips in memory devices and computers; transistors are used in many types of circuits and diodes are used as rectifiers and control devices.

Diodes

Diodes have an anode and a cathode. The cathode end is marked (+) to indicate which polarity it must be connected to in order to work properly. Diodes have many uses. They are found in circuits that are used to change ac to dc for certain control devices. They may also be used as protective devices in circuits. The symbol for a diode is shown in Fig. 3-29.

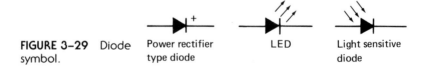

FIGURE 3-29 Diode symbol.   Power rectifier type diode   LED   Light sensitive diode

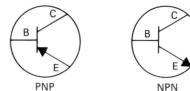

FIGURE 3-30 Transistor symbols, PNP and NPN.

Transistors

The transistor is used for switching and amplification. It consists of three active elements. There are two basic types of transistors, PNP and NPN (see Fig. 3-30). These terms will be discussed later. E stands for emitter, C stands for collector, and B is used to indicate the base connection.

SWITCHES

Switches are used to turn on or off a circuit. They can be made as a simple on-off device or used to control many functions in a number of circuits.

Switches have names that designate what they can do in terms of turning on or off a circuit or circuits. For instance, there are SPST (single pole, single throw) switches, DPDT (double pole, double throw), DPST (double pole, single throw), single-pole, six-position, and so forth. Symbols for various types of switches are shown in Fig. 3-31.

RELAYS

Relays are switches that are moved electrically instead of by hand. They can be made in almost any configuration of poles, throws, and construction. The force that operates a relay is magnetic. The magnetic pull is produced by current passing through a coil of wire. The attraction of an armature causes the switch sections to operate.

Some relay types that you may run into in the air-conditioning and refrigera-

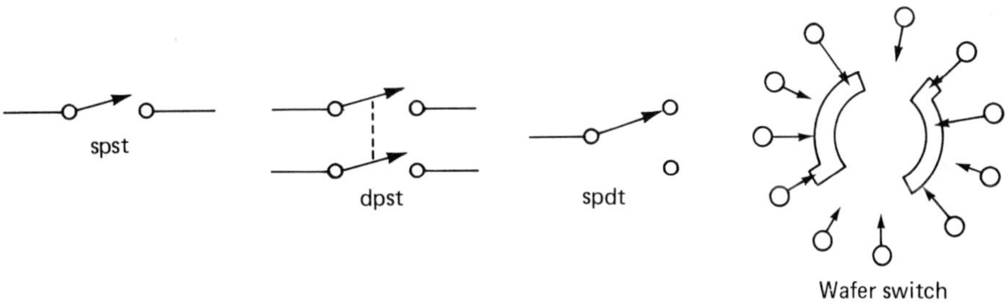

FIGURE 3-31 Switch symbols.

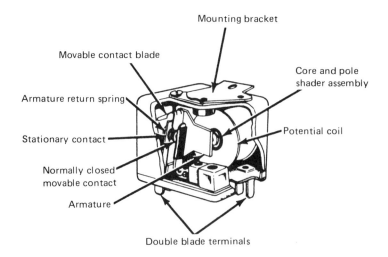

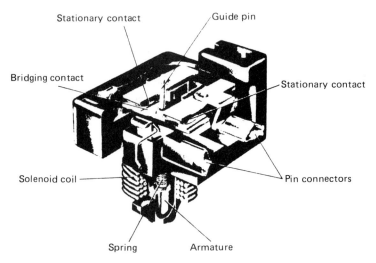

FIGURE 3-32 Relays.

tion field are shown in Fig. 3-32. Relays can also be represented by symbols (see Fig. 3-33).

FUSES AND CIRCUIT BREAKERS

Safety devices are needed on a piece of equipment to protect the expensive equipment and the persons who work on it. The fuse is a safety device designed to protect the equipment from itself. The fuse protects the equipment when it develops a condition that causes it to draw too much current.

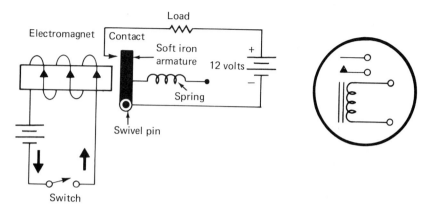

FIGURE 3-33 Relay symbols.

Many circuits need a device to protect the equipment from shorts. Such devices as a fuse or circuit breaker are called for in most electrical circuits. Motors have circuit breakers inserted in their windings that will turn off the motor before it reaches a critically high temperature. Fuses and circuit breakers are represented by a symbol such as that in Fig. 3-34. Typical fuses are shown in Fig. 3-35. They are manufactured in many sizes and values.

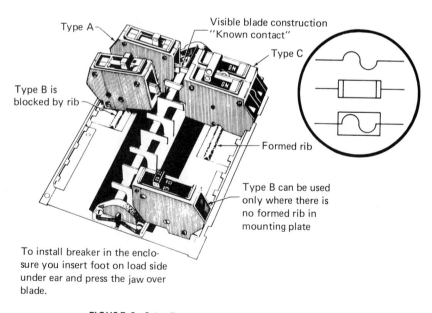

FIGURE 3-34 Fuse symbols and circuit breakers.

62 RESISTORS, COLOR CODE, COMPONENTS, AND SYMBOLS

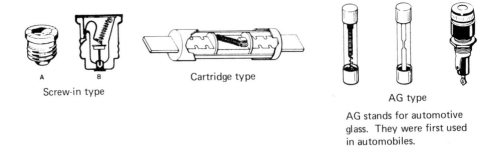

FIGURE 3-35 Different types of fuses.

LAMPS

Indicator lamps come in a variety of shapes and sizes. They may be incandescent, the type used in lamps in the home, or they may be neon types. Light-emitting diodes (LEDs) are also used as indicator lamps. Their symbol is the same as a diode with a couple of arrows pointing away from them to indicate they emit light. However, the incandescent symbol and the neon lamp symbol are quite different (see Figs. 3-36 and 3-37).

Batteries

Many pieces of electronic equipment use batteries. Some types of programmable thermostats use batteries to keep the circuit operating when there is no power. They are required to keep the memory active in the chip so that the settings for times on and off and temperature are not lost with a power failure.

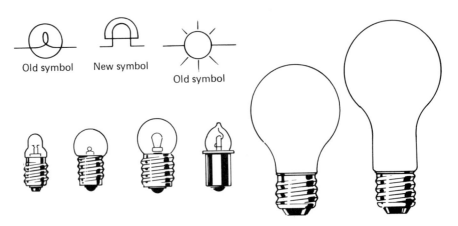

FIGURE 3-36 Incandescent lamps with incandescent-lamp symbols.

METERS 63

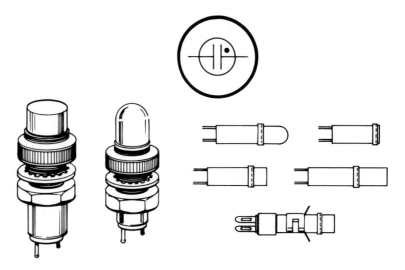

FIGURE 3-37 Neon lamps with neon-lamp symbols.

Batteries are made up of cells. A cell symbol is shown in Fig. 3-38 along with that for a battery. The battery is made up of more than one cell. The way to indicate the voltage is by a number alongside the battery symbol, rather than trying to draw the correct number of cells.

METERS

Many meters are used in electronic and electrical circuits. They are represented by a V or A inside a circle for voltmeter and ammeter, respectively (see Fig. 3-39).

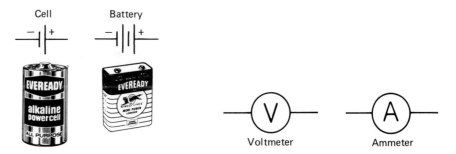

FIGURE 3-38 Cell and battery symbols.

FIGURE 3-39 Meter symbols.

RESISTORS, COLOR CODE, COMPONENTS, AND SYMBOLS

READING SCHEMATICS

As a technician, you will be required to read schematics. This is necessary because all air-conditioning and refrigeration equipment use electrical circuits for operational control. Study this chapter to be sure you can recognize the various symbols used in schematics. The more you study them the easier it will be to interpret circuit meanings.

REVIEW QUESTIONS

1. What is meant by the term resistance?
2. What is a resistor?
3. Why are there so many types of resistors?
4. What is the unit of measurement for resistance?
5. What is the value of a resistor with green, blue, and red for the three stripes on its body?
6. What is meant by tolerance? How does this affect resistors?
7. What is the difference between a potentiometer and a rheostat?
8. What is meant by linear taper?
9. What is a schematic?
10. What are the wattage ratings for carbon-composition resistors?
11. What is a capacitor? Where is it used?
12. How is an electrolytic capacitor different from others?
13. What is an inductor? What does it do?
14. Name two types of semiconductors.
15. What is the difference between a relay and circuit breaker?
16. What is the symbol for a battery?

PERFORMANCE OBJECTIVES

Be able to identify a series circuit.
Be able to identify a fuse and how it is used in a circuit.
Identify a parallel circuit.
Be able to determine the total resistance in a parallel circuit.
Be able to determine the total resistance in a series circuit.
Be able to determine the total resistance in a series-parallel circuit.
Be able to solve problems associated with series, parallel, and series-parallel circuits.

CHAPTER 4

Series and Parallel Circuits

To be able to read a schematic, you have to know the types of circuits used in electrical work. Symbols in series and parallel circuits are the first steps in learning to read a schematic.

SERIES CIRCUIT

A series circuit consists of resistors or other electrical components connected one after another. If you place cells in series, they form a battery. It takes more than one cell to make a battery (see Fig. 4-1). Cells in series produce a higher voltage. Two 1.5-volt cells in series produce 3 volts. The voltages of each cell are added to produce the total voltage. The 9-volt transistor radio battery is actually composed of six 1.5-volt cells connected in series. In a series connection, − is connected to +. The current is limited to whatever is produced by *one* cell. If one cell is dead, the whole battery is useless.

Series circuits have some peculiar properties. They are made up of resistors connected one after the other (see Fig. 4-2). If one resistor is open or removed from the circuit, the whole circuit is open (see Fig. 4-3). In this case, no current flows because the electrons do not have a complete path from − to + of a power source.

Resistances in Series

If the resistances that form a series circuit are added, their total resistance can be found (see Fig. 4-4). When $R_1 + R_2 + R_3 + R_4$ are added, the total resistance in

68 SERIES AND PARALLEL CIRCUITS

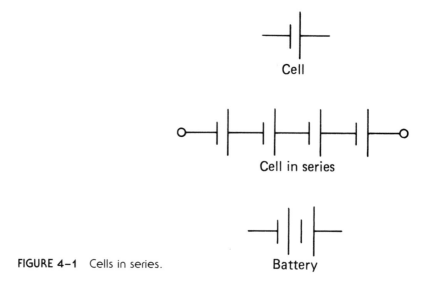

FIGURE 4-1 Cells in series.

FIGURE 4-2 Resistors in series.

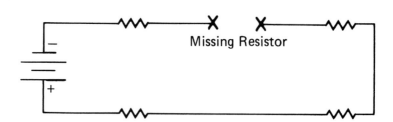

FIGURE 4-3 Series circuit with one resistor missing.

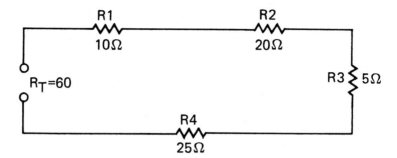

FIGURE 4-4 Total resistance is equal to the sum of the individual resistances.

the circuit is 60 ohms. Or R_1 (10 ohms) + R_2 (20 ohms) + R_3 (5 ohms) + R_4 (25 ohms) is equal to 60 ohms. When this is written as a formula it becomes

$$R_T = R_1 + R_2 + R_3 + R_4$$

Voltages in Series

The total voltage needed for a series circuit is found by adding the voltages needed by each resistor. Or, as Fig. 4-5 shows,

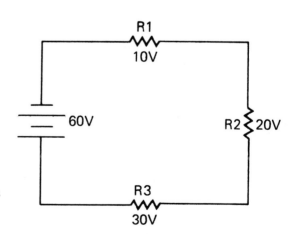

FIGURE 4-5 Voltage drops around the loop equal the applied voltage.

E_A = voltage applied [total voltage]
E_{R1} = voltage drop across R_1
E_{R2} = voltage drop across R_2
E_{R3} = voltage drop across R_3

Step 1: $E_A = E_{R1} + E_{R2} + E_{R3}$
Step 2: $E_A = 10 + 20 + 30$
Step 3: $E_A = 60$ volts

In a series circuit, the voltage applied is equal to the sum of the voltage drops across individual resistors. Voltage in series divides according to the resistance.

Current in Series

Current in a series circuit is the same through all resistors. There is only one path for electrons to move from − to +. If the current in R_1 (Fig. 4-5) is 3 amperes, then the current through R_2 and R_3 is also 3 amperes.

I_T = total current
I_{R1} = current through R_1
I_{R2} = current through R_2
I_{R3} = current through R_3

Therefore, the formula for current in a series circuit is simply stated as

$$I_T = I_{R1} = I_{R2} = I_{R3}$$

Series Circuit Rules

Three basic elements must be considered in any circuit: current, voltage, and resistance. Three simple rules for series circuits concern these elements.

Voltage: Applied voltage is equal to the sum of the individual voltage drops around the series loop.

$$E_A = E_{R1} + E_{R2} + E_{R3}$$

Current: Total current in a series circuit is the same as the current in any resistor in the circuit.

$$I_T = I_{R1} = I_{R2} = I_{R3}$$

SERIES CIRCUIT 71

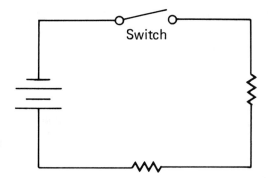

FIGURE 4-6 A switch controls the flow of electrons.

Resistance: Total resistance in a series circuit is equal to the sum of the individual resistances.

$$R_T = R_1 + R_2 + R_3$$

There is a disadvantage in using a series circuit. If one resistor opens, the entire circuit will not operate. An open anywhere in the circuit means the circuit will not function. The path of electrons from − to + terminals of the battery is broken. This feature of a series circuit can be useful. A switch can be used to turn the circuit on and off (see Fig. 4-6).

Fuses

A fuse can also be placed in series to protect the power supply (see Fig. 4-7). If the fuse opens, the circuit is turned off. This means you must locate the trouble. Then you replace the fuse. This safety device (fuse) is very useful.

A fuse is a safety device. Its symbol is ⌒. Fuses are placed in series in a circuit to protect the circuit from excess current. A fuse contains a piece of wire that melts to open the circuit. It melts when too much current flows in the circuit. The size of the fuse determines the amount of current it will safely pass. Fuses are available in a number of sizes and shapes.

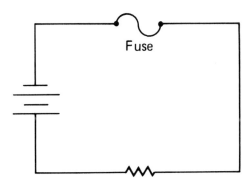

FIGURE 4-7 The fuse is in series with the load.

PARALLEL CIRCUIT

Cells in Parallel

When cells are connected in parallel, the voltage output from the hookup is the same as from one cell. However, the current available is found by adding the current of each cell. Thus, two cells of 1.5 volts each in parallel produces 1.5 volts output. However, if the current available from each of the two cells is 1 ampere, then the total current available is 2 amperes. The type of material used to make a cell determines its voltage. The physical size of a cell determines its current.

Parallel Circuit Characteristics

Parallel circuits are used where a number of devices use the same voltage. This is the case in wiring the lighting circuits in your home (see Fig. 4–8). Most appliances in the home are made to operate on the same voltage, and they are connected in parallel to the same voltage, 120 volts.

A parallel circuit has the same voltage applied to each device. However, the current in a parallel circuit will vary with the resistance of the device.

Currents in a Parallel Circuit

When two resistors are connected in parallel, current can flow through each resistor. Current divides according to the resistance. Two paths for current flow are present. If one path opens, it does not affect the other. The total current in a parallel circuit is equal to the sum of the branch currents, or

$$I_T = I_{R1} + I_{R2} + I_{R3} + \ldots$$

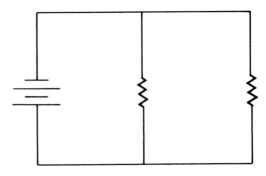

FIGURE 4–8 Two resistors connected across a battery are in parallel.

Resistances in Parallel

Resistances in a parallel circuit cannot be added to yield the total resistance. The following formula must be used to determine the total resistance when more than one resistor is connected in parallel.

$$\frac{1}{R_T} = \frac{1}{R_1} + \frac{1}{R_2} + \frac{1}{R_3} + \cdots$$

For instance, if you have three resistors of 5 ohms, 10 ohms, and 20 ohms in parallel, the total resistance is found as follows:

Step 1: $\dfrac{1}{R_T} = \dfrac{1}{5} + \dfrac{1}{10} + \dfrac{1}{20}$

Step 2: $\dfrac{1}{R_T} = \dfrac{4}{20} + \dfrac{2}{20} + \dfrac{1}{20}$

Step 3: $\dfrac{1}{R_T} = \dfrac{7}{20}$

Step 4: $\dfrac{R_T}{1} = \dfrac{20}{7}$

Step 5: $R_T = 2\dfrac{6}{7}$

A common denominator is necessary to add fractions. Fractions are produced when resistor values are inserted into a formula.

Another formula is used when there are only two resistors in parallel:

$$R_T = \frac{R_1 \times R_2}{R_1 + R_2}$$

For Two Resistors Only. If $R_1 = 10$ ohms and $R_2 = 20$ ohms, then

$$R_T = \frac{10 \times 20}{10 + 20} \quad \text{or} \quad R_T = \frac{200}{30}$$

Thus $R_T = 6\dfrac{2}{3}$ ohms.

Parallel Circuit Rules

Three elements must be dealt with in parallel circuits. They are voltage, current, and resistance. For a quick review, look at these three rules governing parallel circuits:

Voltage: Voltage across each resistor is the same as the applied or total voltage.

$$E_A = E_{R1} = E_{R2} = E_{R3}$$

Current: Total current is equal to the sum of the individual currents.

$$I_T = I_{R1} + I_{R2} + I_{R3} + \cdots$$

Resistance: Total resistance in a parallel circuit is found by using one of two formulas. For two resistors only,

$$R_T = \frac{R_1 \times R_2}{R_1 + R_2}$$

For two or more resistors,

$$\frac{1}{R_T} = \frac{1}{R_1} + \frac{1}{R_2} + \frac{1}{R_3} + \cdots$$

To connect power sources in parallel, make sure the polarity is observed. Connect + to + and − to − to make a parallel hookup.

SERIES–PARALLEL CIRCUITS

Series-parallel circuits use the properties of both series and parallel circuits (see Fig. 4-9). Note that R_1 is in series with R_2 and R_3. R_2 and R_3 are in parallel. Total current flows through R_1 but divides to go through R_2 and R_3. This type of circuit uses the properties of both series and parallel circuits. It is a good idea to make sure you

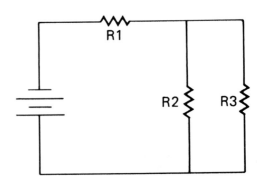

FIGURE 4-9 Series-parallel circuit.

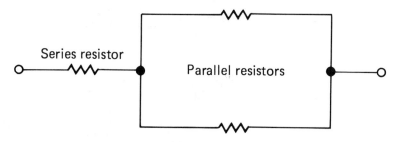

FIGURE 4-10 Series-parallel circuit.

understand the three rules of series circuits and the three rules of parallel circuits before moving ahead.

Series-Parallel Resistance Circuits

It is possible to have circuits that combine series and parallel connections. Within these circuits, some resistors or consuming devices are connected in series. Others are connected in parallel. An example of such a circuit is shown in Fig. 4-10. Another technique for writing a series-parallel circuit is shown in Fig. 4-11. One feature of a series-parallel circuit is obvious from these diagrams: There must be at least three resistors or consuming devices.

Determining Resistance

To determine resistance for a series-parallel circuit, follow a simple method: Calculate the resistance for the parallel units first. Then treat the entire circuit as a series

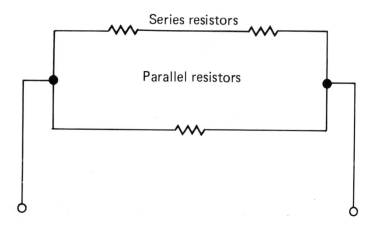

FIGURE 4-11 Alternate series-parallel circuit.

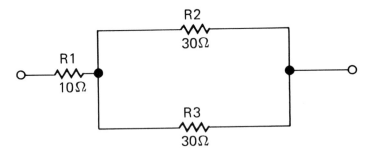

FIGURE 4-12 Series-parallel circuit showing resistance values.

circuit. Thus, *after you have reduced the values of the resistors that are in parallel to a series equivalent,* you add all resistance values together.

For example, look at the diagram in Fig. 4-12. Both parallel resistors have values of 30 ohms. Thus, the parallel circuit has a total resistance of 15 ohms. Remember the rule: *For two resistors of equal value, divide the value of one of the resistors by 2:* 30 ohms divided by 2 = 15 ohms. In effect, you have a series circuit like the one in Fig. 4-13. You might want to redraw the circuit to aid in figuring the values for series-parallel circuit. Figure 4-13 shows two resistors in series. One has a value of 10 ohms. The other has a value of 15 ohms. So the total resistance for the circuit is 25 ohms. Remember the rule: *To determine the total resistance in a series circuit, add the values of all resistors.*

To review these steps for a more complex circuit, consider Fig. 4-14. The diagram shows a circuit with five resistors. To figure the resistance for this circuit, follow the steps shown in Fig. 4-15.

In drawing A of Fig. 4-15, the values of two 20-ohm resistors in series are added together. The total value for these two resistors is 40 ohms. Drawing B then

FIGURE 4-13 Series equivalent of series-parallel circuit.

SERIES–PARALLEL CIRCUITS 77

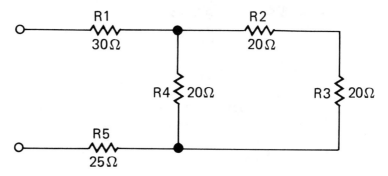

FIGURE 4–14 Schematic of complex series–parallel circuit.

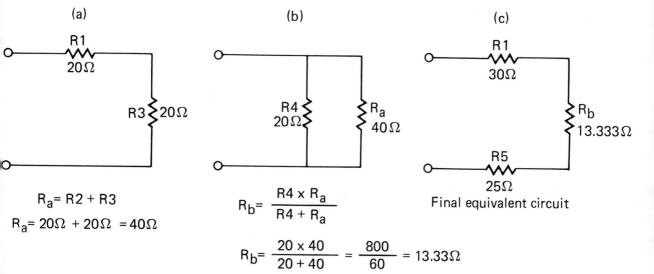

FIGURE 4–15 Redrawn schematics for determining resistance in series–parallel circuit.

shows this value as part of a parallel circuit. This parallel circuit includes the combined values of the two series resistors. These have been redrawn in parallel with a 20-ohm resistor (R_4). The effect is to combine the series resistors within the parallel portion of the circuit from the original drawing. The 20- and 40-ohm parallel resistors have a total resistance of 13.333 ohms. In diagram C, this value is included as part of a series. The three resistors in the final drawing have values of 30, 13.333, and 25 ohms. Adding these, the total resistance for the circuit is 68.333 ohms.

The rules for determining resistance can be summed up as follows:

1. When necessary for clarity, you can redraw the series–parallel circuit.
2. Solve all branches with series resistors first by adding their values.
3. Find the equivalent, or total, resistance of those resistors connected in parallel.
4. After you have series values for all resistors, add all the resistance values.

Determining Current in Series–Parallel Circuits

To determine current in a series–parallel circuit, follow the current path through the entire circuit. Remember your rules of current: In a series circuit, current is the same through all resistors. In a parallel circuit, you divide resistance into voltage to determine current.

Current flow in a series–parallel circuit is shown in Fig. 4-16. The calculations for amperage in this circuit are included in the following section. They are part of the discussion and calculations of voltage values for the same circuit.

Determining Voltage in Series–Parallel Circuits

To find voltage drop across resistors in series–parallel circuits, follow the rules you already know. When resistors are in series, *add* voltages. When resistors are in parallel, voltage is the *same* across all resistors.

Examine Fig. 4-16 and see how the current divides; then look at Fig. 4-17. This figure provides total voltage at the source and resistance values for all resistors. To find values for voltage and current, here are the simple steps to follow:

1. Redraw the circuit for clarity, if necessary. Follow the steps illustrated in Figs. 4-18, 4-19, and 4-20.
2. Locate all branches with series resistors. This has been done in drawings 1 and 2 of Fig. 4-18.
3. Locate all branches with parallel resistors. This has been done in drawing 3 of Fig. 4-19.
4. Locate all branches that have series resistors in series with other resistors. Look back at Fig. 4-17. Resistors 1, 2, and 5 are in series. These are shown in series in loop 1 in Fig. 4-19. Resistors 1, 3, 4, and 5 are also in series. These are shown in loop 2 in Fig. 4-20.
5. Apply the rules for series or parallel circuits to calculate voltage and current.

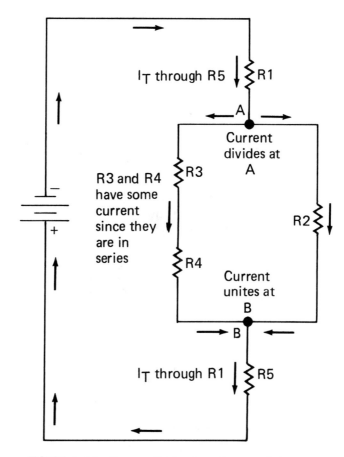

FIGURE 4-16 Current division in series–parallel circuit.

To figure current, divide total resistance into total voltage. Total voltage is given as 100 volts. Resistance values are shown in Fig. 4-17. Look at the resistance for the parallel portion of this circuit first. Resistors 3 and 4 are in series. These values of 10 ohms each are added for a total of 20 ohms. This gives the effect of two 20-ohm resistors in parallel. Two 20-ohm resistors in parallel give a total resistance value of 10 ohms. This value (10 ohms) is then added in series to the values of resistors 1 and 5. The values added are 5 ohms, 10 ohms, and 10 ohms, for a total of 25 ohms. Dividing 25 ohms into 100 volts (using Ohm's law) produces an answer of 4 amperes.

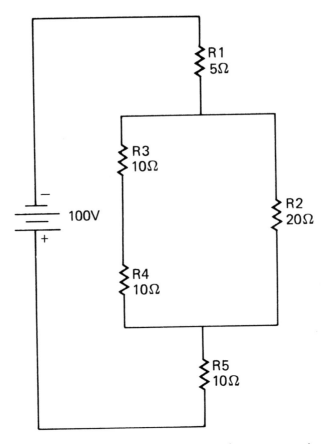

FIGURE 4-17 Complex series–parallel circuit with resistance values given.

$$R_3 + R_4 = 10 \ \Omega + 10 \ \Omega = 20 \ \Omega$$

$$20 \ \Omega \div 2 = 10 \ \Omega$$

$$\text{resistance of parallel circuit} = 10 \ \Omega$$

$$10 \ \Omega + R_1 + R_3 = 10 \ \Omega + 5 \ \Omega + 10 \ \Omega = 25 \ \Omega$$

Total resistance for the circuit is 25 ohms. Now, to determine current, divide the voltage (100 V) by the resistance.

$$100 \ \text{V} \div 25 \ \Omega = 4 \ \text{A}$$

SERIES–PARALLEL CIRCUITS   81

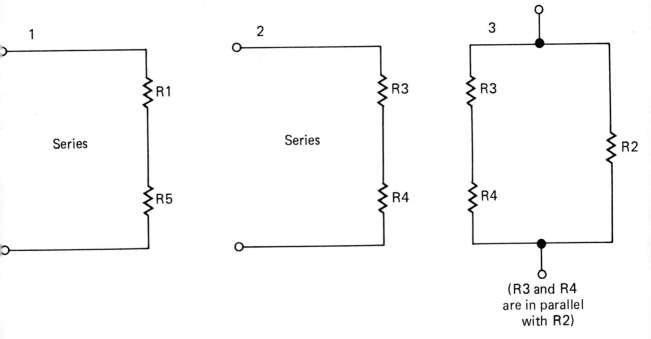

FIGURE 4-18 Complex series–parallel circuit redrawn in three branches.

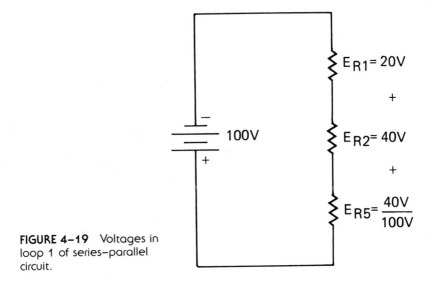

FIGURE 4-19 Voltages in loop 1 of series–parallel circuit.

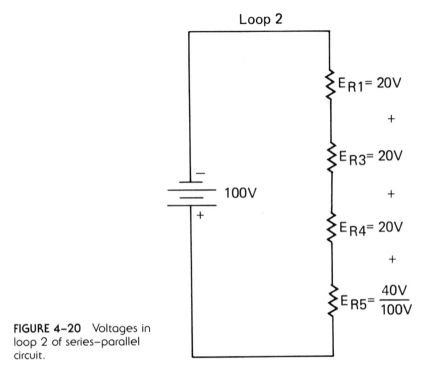

FIGURE 4-20 Voltages in loop 2 of series–parallel circuit.

The current through the circuit is 4 amperes. This current is divided through the parallel portion of the circuit. So current at resistors 2, 3, and 4 is 2 amperes. At resistors 1 and 5, the current is 4 amperes.

Note that resistors 1 and 5 are repeated in the series drawings in Figs. 4-19 and 4-20. This is important to remember: In determining voltage in a series–parallel circuit, reduce the branches to an equivalent series value. In doing so, consider all resistors in each series.

To determine voltage for each resistor in a series branch, multiply the current by the resistance (Ohm's law: $E = I \times R$). Resistor 1 has a rating of 5 ohms and a current of 4 amperes. Use Ohm's law again and get $4 \times 5 = 20$. The voltage is 20. Resistor 2 has 20 ohms of resistance and a current of 2 amperes. Using Ohm's law again, you get 40 volts for the voltage drop across R_2. Resistor 5 (R_5) has 10 ohms with 4 amperes of current. This produces 40 volts for R_5. R_3 and R_4 each have 10 ohms resistance. These resistors have 20 volts across each.

You have been working with series, parallel, and series–parallel circuits. They are the basic circuits of anything electrical or electronic. Seeing how voltage, current, and resistance behave in these circuits brings about an understanding of circuits and increases your ability to read schematics. Schematics are symbols placed in series or parallel configurations. They are the shorthand of the electrical field. Every technician should be able to read a schematic and be able to identify how the components are connected in a circuit.

REVIEW QUESTIONS

1. What is a series circuit?
2. What is a fuse? How is it used?
3. What is a parallel circuit?
4. What is the formula for finding total resistance in parallel?
5. What is the formula for finding total resistance in series?
6. How do you determine current in a parallel circuit?
7. What is a series-parallel circuit?
8. How is total resistance found in a series-parallel circuit?

PERFORMANCE OBJECTIVES

Know the basics of magnetic induction.
Know how electromagnets are made.
Know how a solenoid differs from a relay.
Know the primary purpose of an electrically operated solenoid valve.
Know the difference between NC and NO.
Know what happens when a valve leaks hot gas in a hot-gas defrost system.
Know the usual voltage ratings for solenoid valves.
Know what VA stands for and why you need to know it.

CHAPTER 5

Magnetism, Solenoids, and Relays

Magnetism is something everyone knows about and has worked with, even when very young. Magnetism is *the property of certain materials that permits them to produce or conduct magnetic lines of force.*

Magnetic lines of force surround a current-carrying conductor or any atomic particles in motion. Even in the atom, magnetic fields result from the spinning motions of the electrons. Such magnetic lines of force can also interact with electric fields or other magnetic fields.

Everytime you press a doorbell button, an electromagnet causes it to ring or chime. Electromagnets are created by sending electricity through a coil.

Permanent magnets are made from materials that will retain magnetism. That is, they keep their ability to attract iron (and other materials). They do not need electricity to function. Electromagnets lose their magnetism when the electric current is removed. Both permanent and temporary (electromagnet) magnets have their use in heating, air-conditioning, and refrigeration control circuits.

PERMANENT MAGNETS

One type of permanent magnet is the natural magnet, the lodestone. It is also possible to make permanent magnets. For example, if you stroke a piece of high-carbon steel with a lodestone, the steel becomes magnetized (see Fig. 5-1).

Under normal conditions, the steel stroked with the lodestone keeps its magnetism permanently. When this happens, the molecules that form the steel bar line up in the same direction (see Fig. 5-2). The molecules that line themselves up in this

86 MAGNETISM, SOLENOIDS, AND RELAYS

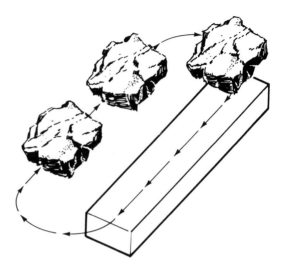

FIGURE 5-1 Making a permanent magnet by rubbing it with a lodestone.

way point toward the ends, or poles, of the magnet. Each magnet has two poles, which are identified as north and south.

A permanent magnet can be used to create other magnets. Figure 5-3 shows a permanent magnet used to stroke a bar of high-carbon steel. This creates a second magnet.

Magnets made of high-carbon steel sometimes lose their magnetism. This happens if the magnetized steel bar receives a strong shock, for instance, if the bar is dropped or struck with a hammer. When this happens, the molecules lose their alignment. This leads to a loss of magnetism. Magnetism can also be lost if a magnet is heated to high temperatures.

A magnet may be formed by rubbing a piece of steel with another magnet. This is called *magnetic induction*. This was the only way known of making magnets until electromagnetism was discovered.

TEMPORARY MAGNETS

A bar of low-carbon steel attracts iron particles when it is in direct contact with another magnet. For example, you can place a lodestone against one pole of a bar

Unmagnetized steel

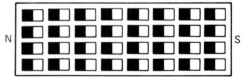
Magnetized steel

FIGURE 5-2 A magnetic field aligns iron molecules to produce a permanent magnet.

MAGNETIC THEORY

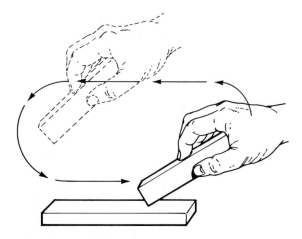

FIGURE 5-3 Permanent magnet used to make another permanent magnet.

of low-carbon steel as shown in Fig. 5-4. When this is done, the bar attracts iron particles. Removing the lodestone removes the magnetism. Then the iron particles fall away. The same happens if you place a permanent magnet against one end of a bar of low-carbon steel.

Low-carbon steel does not retain, or keep, magnetism. This material can serve as a temporary magnet. But it cannot be used for a permanent magnet.

ELECTROMAGNETS

Magnetism also results when electric currents are passed through iron materials. Electromagnetism is important to electricity.

MAGNETIC THEORY

It is important to know the basics about how magnetism behaves for you to understand how certain pieces of equipment operate. Scientists still do not know exactly

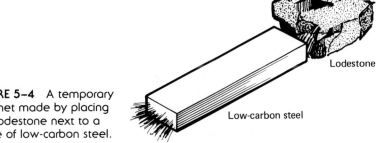

FIGURE 5-4 A temporary magnet made by placing the lodestone next to a piece of low-carbon steel.

how magnets behave. There are no laws or fixed rules for magnetism. Instead, there are a number of theories. These theories explain the behavior patterns that can be observed.

Magnetic Permeability

The ability of a material to become magnetized is called *permeability*. The magnetic force that attracts iron materials is called *magnetic flux*. The greater the permeability of a material, the higher its magnetic flux.

Numbers are used to indicate the permeability of materials. The higher the number, the more easily a material is magnetized. The relative permeabilities of a number of materials are listed in Table 5-1. At lower levels, materials are considered to be nonmagnetic. This applies to aluminum and all materials listed above it in Table 5-1.

Materials with higher ratings, including nickel, cobalt, iron, and permalloy, have high permeability. Permalloy, a combination of materials, has extremely high permeability. When a bar of permalloy is held in a north–south direction, it becomes a magnet. When the bar is turned to face east and west, it loses the magnetism.

Shapes of Magnets

Magnets can be made in a variety of shapes. Some of these are shown in Fig. 5-5. One common shape for magnets is the rod or bar. Another shape is similar to the letter U. This is sometimes called a horseshoe magnet.

Magnets made in a horseshoe shape have an advantage: Their ends, or poles, are close together. Thus, horseshoe magnets have a strong field of magnetic attraction, or flux field (see Fig. 5-6). By comparison, a bar magnet of the same strength has a flux field with less attraction. This is demonstrated in Fig. 5-7. The horseshoe magnet comes in handy when we start studying electrical measurement instruments.

TABLE 5-1 Selected Permeabilities

Material	Permeability
Bismuth	0.999833
Quartz	0.999985
Water	0.999991
Copper	0.999995
Liquid oxygen	1.00346
Oxygen (STP)[a]	1.0000018
Aluminum	1.0000214
Air (STP)	1.0000004
Nickel	40.
Cobalt	50.
Iron	7,000.
Permalloy	74,000.

[a]STP, standard temperature and pressure.

MAGNETIC THEORY 89

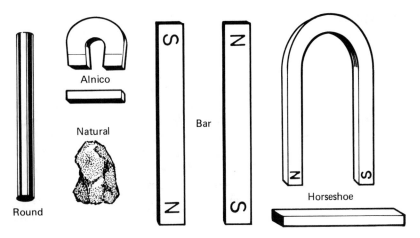

FIGURE 5-5 Various shapes of magnets.

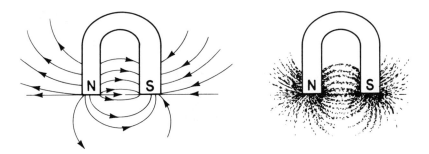

FIGURE 5-6 Horseshoe-magnet flux pattern.

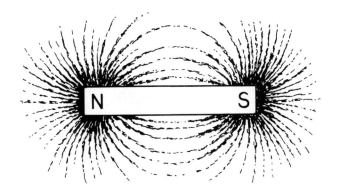

FIGURE 5-7 Bar-magnet flux field.

Poles of Magnets

Dip a bar magnet into a pile of iron filings. When it is withdrawn, the filings will gather at the ends of the magnet (see Fig. 5-8). This demonstrates that magnetism is strongest at the ends. These are the poles of a magnet.

Each magnet has two poles. However, there is a difference between the poles. If you suspend a magnet from a string, it turns until one pole points north and the other south (see Fig. 5-9). (This will occur if you do not have any large deposits of steel around, such as steel storage cabinets or a building with a steel framework.) This is the principle behind the operation of a compass: One pole of a magnet is north-seeking. A bar magnet is suspended on a bearing to reduce friction. The north-seeking pole points north and the other pole is direct south. Therefore, the poles of a magnet are referred to as the north pole and the south pole.

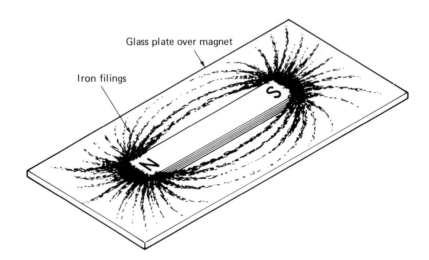

FIGURE 5-8 Magnetic flux is strongest at the poles.

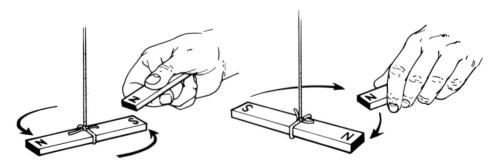

FIGURE 5-9 Magnetic poles align to north and south.

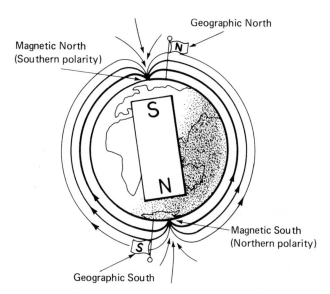

FIGURE 5-10 Polar attraction of magnets.

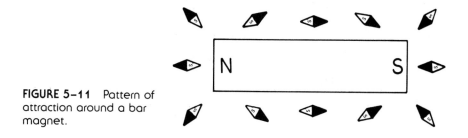

FIGURE 5-11 Pattern of attraction around a bar magnet.

If you point the north poles of two bar magnets at each other, they are *not* attracted. Actually, there is a reverse force pushing the magnets apart. This force repels them from each other. This is a rule of magnetism: *Like poles repel.*

If you place the north pole of one magnet near the south pole of another, there is a strong attraction. This follows another rule of magnetism: *Unlike, or opposite, poles attract.* This means that the north-seeking pole of a magnet is actually the south pole (see Fig. 5-10). The pattern of attraction around a bar magnet is shown in Fig. 5-11.

ELECTROMAGNETISM

The relationship between electric current and magnetism was discovered by Hans Christian Oersted in 1820. This relationship is direct. When current flows through a conductor, there is a magnetic field around the conductor. The direction of the

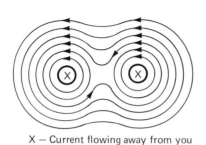

X — Current flowing away from you

FIGURE 5-12 Current flow creates an electromagnetic field.

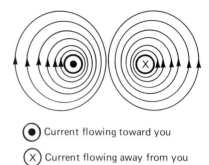

⊙ Current flowing toward you

⊗ Current flowing away from you

FIGURE 5-13 Currents in opposite directions create repelling force fields.

current flow determines the force field of an electromagnet. Currents flowing in the same direction set up connecting fields of force (see Fig. 5-12). Currents flowing in opposite directions set up repelling force fields (see Fig. 5-13).

The polarity of an electromagnet is determined by the direction of the current (see Fig. 5-14). This is important in an electrical meter.

The strength of an electromagnetic field is proportional to the current flowing through the conductor. More current means more magnetism.

Magnetism in a Coil of Wire

The magnetic field that surrounds a conductor depends on the form into which the wire is shaped (see Fig. 5-15). The magnetic field surrounding a single loop of wire is shown in Fig. 5-15A. Additional loops are shown in Fig. 5-15B. These loops form a spiral or *helix*. This shape is also known as a helical coil. Figure 5-15B shows magnetic flux around loops of wire wound next to one another. The more loops in the wire, the stronger the magnetic field becomes. This is shown in Fig. 5-15C.

The strength of an electromagnet or coiled conductor depends on the amount of current flowing in the coil and the number of turns of wire. Keep in mind that the direction of current flow determines the magnetic polarity (see Figs. 5-13 and 5-14).

ELECTROMAGNETS

The usual form of an electromagnet is a coil of wire wound around a soft iron core. Thus, the core will lose its magnetism when the source of energy is removed (see Fig. 5-16). The iron core provides an easy path for the magnetic field created by the coil. An electromagnet with an iron core forms a stronger magnet than the same coil of wire without the iron core.

The size of the iron core can help determine the strength of an electromagnet. For full strength, the core should be large enough to absorb all the magnetism from

ELECTROMAGNETS

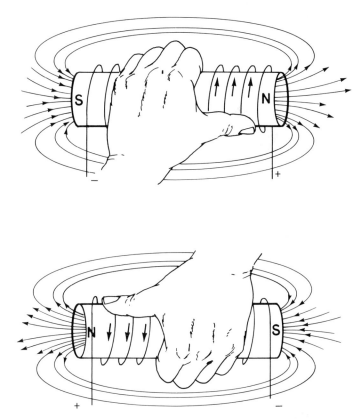

FIGURE 5-14 The left-hand rule shows how the polarity of an electromagnet is determined by the current flow. Grasp the coil with your fingers pointing in the direction of the current flow; the thumb of the left hand then points toward the north pole.

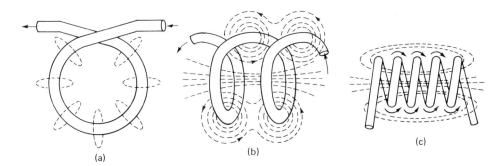

FIGURE 5-15 Magnetic fields form around various shapes of magnetic coils.

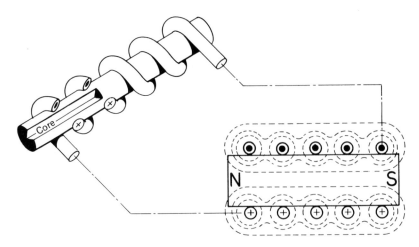

FIGURE 5-16 Electromagnets can be made by winding wire around a soft iron core.

the coil. When the coil creates more magnetism than the core can absorb, it is called *saturation*. If the core is too small, there is too much magnetism to be absorbed. All the magnetism of the coil can be used when the core has a capacity slightly larger than the magnetic flux created.

Using Electromagnetism

The generation and use of electricity are directly related to electromagnetism. Some of these uses are in the solenoid and relay. These two items are useful in heating, air-conditioning, and refrigeration equipment control circuits. It is important to understand how they work, since they can be the source of much troubleshooting time.

THE SOLENOID

A magnetic field seeks the path of minimum *reluctance,* just as an electric current seeks the path of least resistance. Reluctance and resistance are related. Resistance refers to the opposition to current flow in a circuit. Reluctance refers to the opposition of the flow of a magnetic field. The lower the reluctance, the greater is the attraction of materials to a magnetic field.

A solenoid is a device that uses these principles. A helical coil of wire produces a magnetic field. An iron core fits loosely within the coil of wire. When the current is off, the core rests outside the area of the coil. When the current is applied, the core is sucked into the coil. This is referred to as the *sucking effect* of a coil (see Fig. 5-17).

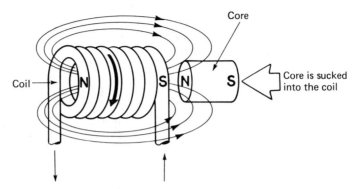

FIGURE 5-17 The *sucking effect* of a coil.

The sucking effect is often used in devices that require a small amount of physical movement. One type of solenoid is shown in Fig. 5-18. A common use for solenoids is in devices called solenoid valves (see Fig. 5-19). This valve is a device that opens and closes to permit liquids or gases to flow. In most valves, for example, you turn a handle or knob to start or stop the flow of gas. Solenoid valves are used widely as safety devices. They are found in gas lines, air lines, and water lines. Many types are used in air-conditioning, refrigeration, and heating systems.

The solenoid in the valve in Fig. 5-19 is closed when there is no current through its coil. The valve stem is held in a closed position by a spring that applies light pressure. When the valve is closed, gas cannot flow. Current is applied when a person or thermostat turns on a heater. The current draws the movable core, or plunger, into the coil. This opens the gate to the flow of gas. When the current is turned off, the spring moves the valve system back into a closed position.

This same principle is used widely in solenoid-type relays. A solenoid relay is like an electrically operated switch. The coil in this type of relay pulls a core piece that has a number of electrical contacts. The contacts are designed so that they may either close or open electrical circuits. The relay contacts themselves can be designed to handle large amounts of current. But the coil of the control solenoid may operate on only a fraction of an ampere. The effect is that low-voltage, low-current electricity is used to control the flow of larger amounts of electricity.

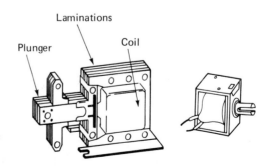

FIGURE 5-18 Two types of solenoids.

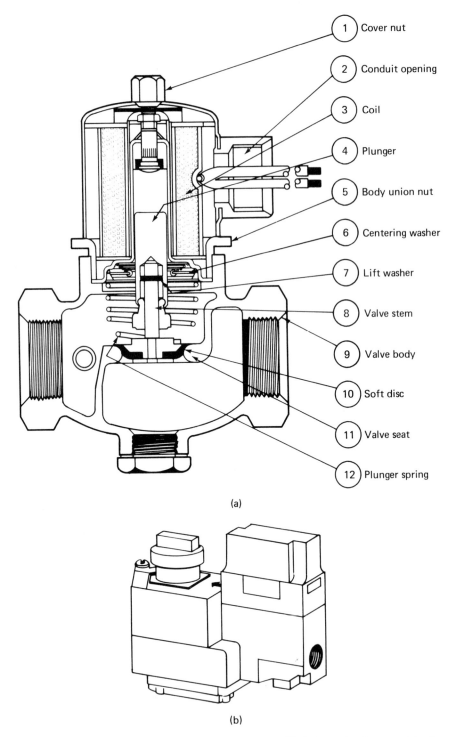

FIGURE 5-19 (A) Cutaway view of a solenoid valve. (B) Solenoid valve used for hot-air furnace gas control. (C) Solenoid used to operate a relay. (D) Relay used to start a refrigerator compressor.

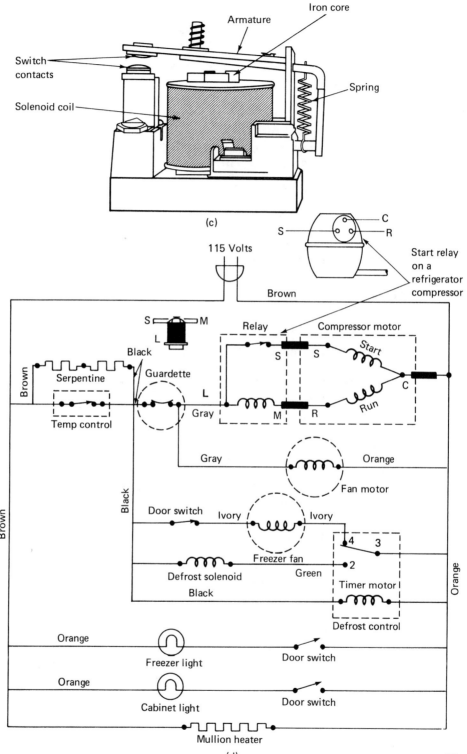

Power Relays

A heavy-duty power relay that operates from solenoid action is shown in Fig. 5-20. In this device, a spring pulls the solenoid core, or armature, away from the electrical contact when the current is off. When current flows, the armature is pulled toward the coil. An electrical contact connected to the armature is either closed or opened by this action. Note that the electromagnetism can be used either to open or close the relay. The action taken depends on the design of the relay. The relay contacts can be arranged for a variety of functions, such as SPST, SPDT, DPDT, or any other combination.

The advantage of a relay is that a substantial pulling power can be developed with a small coil current. The contacts can be made quite large and can handle or switch high values of electrical power. An extremely small amount of control power thus can be used to switch much higher voltages and currents in a safe manner (see Fig. 5-21).

Solenoid Valves

The primary purpose of an electrically operated solenoid valve is to control automatically the flow of fluids, liquid or gas. There are two basic types of solenoid valves. The most common is the normally closed (NC) type, in which the valve opens when the coil is energized and closes when the coil is de-energized. The other type is the normally open (NO) valve, which opens when the coil is de-energized and closes when the coil is energized (see Fig. 5-22).

Principles of Operation

Solenoid valve operation is based on the theory of the electromagnet. The solenoid valve coil sets up a magnetic field when electrical current is flowing through it. If a magnetic metal, such as iron or steel, is introduced into the magnetic field, the pull of the field will raise the metal and center it in the hollow core of the coil. By attaching a stem to the magnetic metal plunger, this principle is used to open the port of the valve. When the electrical circuit to the coil is broken, the magnetic field

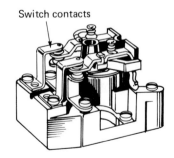

FIGURE 5-20 Heavy-duty power relay.

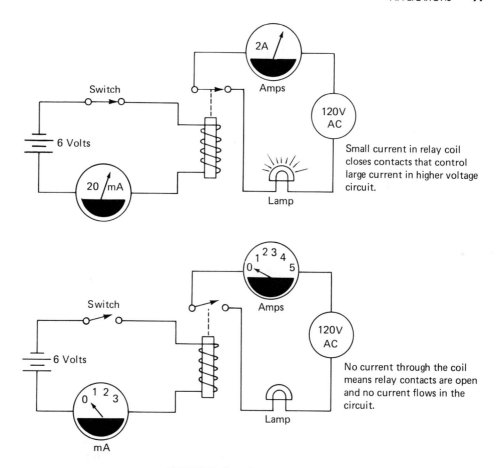

FIGURE 5-21 Relay in a circuit.

collapses and the stem and plunger either fall by gravity or are pushed down by the kick-off spring.

Some solenoid valves are designed with a hammer-blow effect. When the coil is energized, the plunger starts upward before the stem. The plunger then picks up the stem by making contact with a collar at the top. The momentum of the plunger assists in opening the valve against the unbalanced pressure across the port.

APPLICATIONS

In many cases, valves are used for controlling the flow of refrigerants in liquid or suction lines or in hot-gas defrost circuits. They are equally suitable for many other less common forms of refrigerant control.

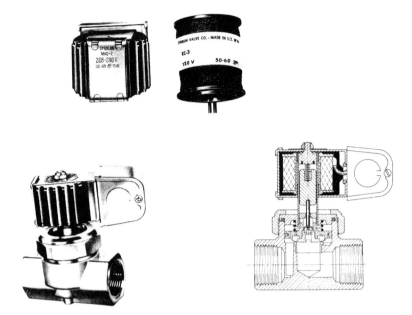

FIGURE 5-22 Solenoid coil and step-down transformer. (Courtesy of Sporlan)

Liquid Line Service

The primary purpose of a solenoid valve in a refrigerant liquid line is to prevent flow into the evaporator during the off cycle. On multiple systems, a solenoid valve may be used in each liquid line leading to the individual evaporators.

The application of a liquid line solenoid valve depends mainly on the method of wiring the valve with the compressor control circuit. It may be wired so the valve is energized only when the compressor is running. This type of application is shown in Fig. 5-23.

Another application, known as pump-down control, uses a thermostat to control the solenoid valve (see Fig. 5-24 for a wiring and valve location schematic). When the thermostat is satisfied, the valve closes and the compressor continues to run until a substantial portion of the refrigerant has been pumped from the evaporator. A low-pressure cutout control is used to stop the compressor at a predetermined evaporator pressure. When the thermostat again calls for refrigeration, the solenoid valve opens, causing the evaporator pressure to rise and the compressor to start. This arrangement can be used on either single or multiple evaporators.

Suction Line Service

There are several applications, particularly on suction lines, where pressure drops in the range of 2 to 4 pounds per square inch (psi) cannot be tolerated. Therefore,

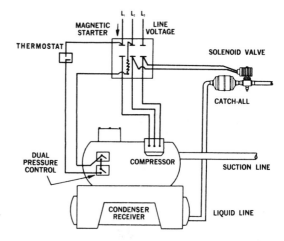

FIGURE 5-23 Liquid-line solenoid valve application. (Courtesy of Sporlan)

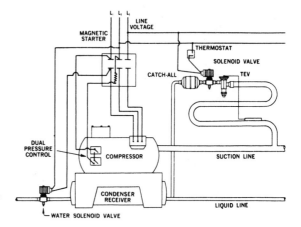

FIGURE 5-24 A pump-down control uses a thermostat to control the solenoid valve. (Courtesy of Sporlan)

only valves that are capable of opening at very low pressure drops are suitable for this type of use. Some valves incorporate the floating disc principle and are ideally suited for such special applications. They are capable of opening full at pressure drops of 0.1 psi and below.

Larger-capacity valves are suitable for suction service when supplied with internal parts that are mechanically connected. With this arrangement, the piston is connected to the stem and plunger assembly, and when the coil is energized, the plunger assists in supporting the piston. As a result, the pressure drop through the valve is reduced to a bare minimum. Valves with the direct-connected assembly are designated usually with a prefix D to the type number.

When these valves are required for suction service, they are supplied with a spring support under the piston. The spring counterbalances the major portion of the piston's weight, and therefore the valve will open with far less pressure drop than normal. Valves with the counterbalance spring are identified by the prefix S added to the type number.

HIGH-TEMPERATURE APPLICATIONS

In some high-temperature applications, a high-temperature coil construction is required. The temperature of the fluid or gas flowing through the solenoid valve will generally determine whether a high-temperature coil is necessary.

HOT-GAS DEFROST SERVICE

Several piping arrangements are used for hot-gas defrost systems, one of which is shown in Fig. 5-25. A portion of the compressor discharge gas is passed through the solenoid valve into the evaporator. The solenoid valve may be controlled either manually or automatically for this duty.

Hot-gas defrost valve selection requires a slightly different approach from the simple pressure drop versus tonnage. Be sure to consider the evaporator temperature correction factors to make certain that the valve selected has adequate capacity.

Normally open solenoid valves have many uses. Perhaps the most popular is their adaptation to heat-reclaiming systems. The use of one normally closed valve and one normally open valve to shunt the discharge gas to either the outdoor condenser or the indoor heat-reclaiming coil provides positive opening and closing action.

This eliminates the problem found in some three-way valves which have a tendency to leak hot gas into the heat-reclaiming coil when not required. When this leakage occurs during the cooling season, it imposes an extra load on the cooling system that wastes energy, rather than conserving it.

If leakage occurs during the heating season when all the discharge gas should

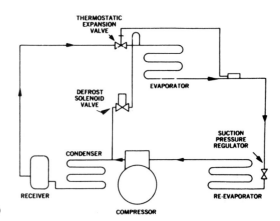

FIGURE 5-25 Solenoid valves used for hot-gas defrost. (Courtesy of Sporlan)

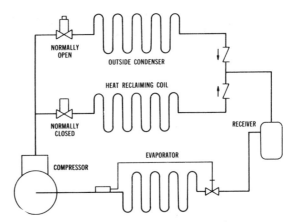

FIGURE 5-26 Heat-reclaiming cycle schematic. (Courtesy of Sporlan)

be going to the reheat coil, a good portion of the liquid charge could become logged in the inactive condenser. For a simple schematic of a heat-reclaiming cycle, see Fig. 5-26. Many original equipment manufacturers (OEMs) have developed their own reheat cycle, which may be completely different from the one illustrated. In addition, some may incorporate head pressure control as well, so it is always advisable to consult the manufacturer's bulletin regarding its particular design.

TRANSFORMERS FOR LOW-VOLTAGE CONTROLS

The use of low-voltage control systems is becoming more widespread as the demand for residential air conditioning increases. This necessitates the use of a transformer for voltage reduction, normally to 24 volts. The selection of a transformer is not accomplished by merely selecting one that has the proper voltage requirements. The volt–ampere (VA) rating is equally important. To determine the VA requirements for a specific solenoid valve, refer to the manufacturer's data. Insufficient transformer capacity will result in reduced operating power or a lowering of the maximum operating pressure differential (MOPD) of the valve. MOPD and safe working pressure (SWP) are usually noted on the valve nameplate.

FIGURE 5-27 Valve nameplates with voltage, wattage, frequency, MOPD, and SWP. (Courtesy of Sporlan).

If more than one solenoid valve and/or other accessories are operated from the same transformer, the transformer VA rating must be determined by adding the individual accessories' VA requirements.

REVIEW QUESTIONS

1. What is magnetism?
2. What is magnetic induction?
3. How is an electromagnet made?
4. What is a solenoid? Where is it used?
5. What is a relay?
6. How is a relay different from a solenoid?
7. What is meant by like poles repel and unlike poles attract?
8. Where are electromagnets used?
9. What is the primary purpose of an electrically operated solenoid valve?
10. What do NO and NC mean?
11. How are valves with a counterbalance spring identified?
12. What happens when a valve leaks hot gas in a hot-gas defrost system?
13. What is the usual voltage rating for solenoid valves?
14. What does VA mean?
15. Why do you need to know the VA rating of a transformer?

PERFORMANCE OBJECTIVES

Know how electricity is measured.

Know how the d'Arsonval meter movement works.

Know what is meant by a meter's full-scale deflection.

Know how to correct for parallax error.

Know the difference between *milliampere, microampere,* and *ampere.*

Know how a meter's range is extended.

Know how an ammeter is connected in a circuit.

Know the difference between and uses for series and shunt ohmmeters.

Know how to use a megger.

Know how to use the voltmeter to test a circuit.

CHAPTER 6

Electrical Measuring Instruments

Electricity cannot be seen, heard, tasted, or smelled. That is why it is necessary to use some type of device to detect its presence. Once you have detected its presence, you need to know how much of it is present. This is the function of electrical measuring instruments. Another reason for using an instrument is to allow you to detect the presence of electricity without having to *feel* it. The sense of feel is the only one of our five senses that can be used to detect the presence of electricity. This can be deadly also; therefore, a measuring device is much more reliable and safer.

It takes a meter to measure electricity. It takes a meter to determine the three properties of a circuit: voltage, current, and resistance. The main reason for using meters in heating, refrigeration, and air-conditioning work is to aid in troubleshooting defective equipment. With this in mind, we will examine a number of types of meters. By looking a little closer at what is under the meter cover, you will better understand how to treat the device to make it last longer and produce the accuracy needed to do your job.

The most frequently used meter is the multimeter (see Fig. 6-1). It can be used to measure voltage, current, and resistance.

TYPES OF METER MOVEMENTS

There are many types of meter movements. One of the most often used in the analog configuration is the d'Arsonval meter movement, named after its designer, Arsene d'Arsonval. Actually, it is a revision of the galvanometer. The galvanometer is a measuring device that detects the presence and direction of current flow in a circuit.

108 ELECTRICAL MEASURING INSTRUMENTS

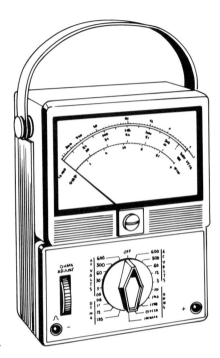

FIGURE 6-1 Multimeter.

Edward Weston made some improvements on the design that d'Arsonval produced and the results are still seen in today's meters.

D'Arsonval Meter Movement

The d'Arsonval meter movement has a small, rectangular coil of wire. The coil is suspended in a magnetic field created by a permanent magnet. When current is applied to the coil, it becomes an electromagnet. The energized coil then lines up with the poles of the permanent magnet. The amount of current applied to the coil controls its movement. An electromagnet that is free to move will align its axis with the magnetic axis of a fixed magnet. The axis of an electromagnet is a straight line between its poles (see Fig. 6-2).

The coil must be free to rotate in order to align itself with the magnetic axis. The coil in the meter is mounted on pivots that permit easy rotation. Two small springs are mounted on the top and bottom. These springs offer slight resistance to the rotation of the coil. The springs control the position of the coil when there is no current flowing.

When current flows in the coil, it produces a magnetic field around the coil. This magnetic flux overcomes the force of the springs and moves the coil. A pointer on top of the coil rotates to mark the amount of movement. The greater the current through the coil, the more it turns—and the further the pointer moves. The pointer

TYPES OF METER MOVEMENTS

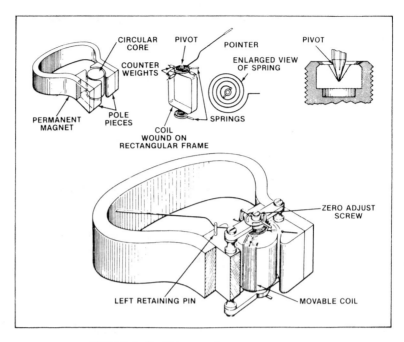

FIGURE 6-2 D'Arsonval meter movement.

then stops in front of a marked scale on the face of the meter. The position of the pointer indicates the meter measurement (see Fig. 6-3).

A typical meter scale is shown in Fig. 6-4. Note the even divisions between each line on the scale. This is a linear-type meter. Every space indicates the same amount of voltage.

As you look closer at the d'Arsonval meter movement, you will find that it can be extremely sensitive to small currents. For instance, a relatively inexpensive

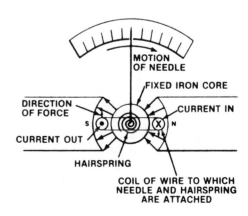

FIGURE 6-3 D'Arsonval meter movement with scale.

110 ELECTRICAL MEASURING INSTRUMENTS

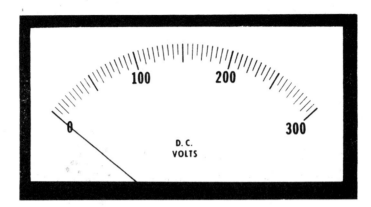

FIGURE 6-4 DC-volts meter scale.

meter can give a full-scale deflection of 10 microamperes (μA). *Full-scale deflection* is the total range of a meter scale. A microampere is one-millionth (0.000001) of an ampere.

Meter deflection is determined by three factors:

1. The number of turns of wire in the coil and the amount of current flowing in the coil.
2. The strength of the magnetic field produced by the permanent magnet affects the positioning of the coil.
3. The tension of the springs and the friction of the bearings determine the sensitivity of the meter movement.

The amount of current flowing through the meter determines its pointer deflection since it is the only variable in the meter movement circuit. The scale of the meter is calibrated (marked) to show the type of reading (volts, ohms, or amperes). The scale in Fig. 6-4 is calibrated to read dc (direct current) volts.

Before you use a meter, check its instruction manual. Each meter is a little different and you should be aware of any important instructions in the instruction manual.

Analog Meters

An analog-type meter has a scale that uses a pointer to indicate its reading. The digital meter uses digits similar to a digital watch to indicate the voltage, current, or resistance. There is a place for both types. At this point, it is best to concentrate on the analog type since many of them are still in use and they are still made for all types of applications.

A scale with four separate readings is shown in Fig. 6-5. This shows what is known as a *linear* scale. A linear scale has marks spaced equally. The full-scale deflection is 5. This could mean 5 amperes, 5 volts, or 5 milliamperes, or 5 of any

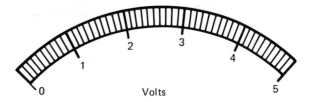

FIGURE 6-5 Linear scale.

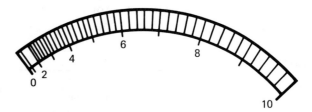

FIGURE 6-6 Nonlinear scale.

other value of voltage or current. The information on what is being measured usually appears below the scale. In this case (Fig. 6-5) it is volts.

A *nonlinear* scale is shown in Fig. 6-6. This particular scale is referred to technically as a *square-law scale*. The lines in this scale increase in squares. That is, each larger scale marking indicates that the measured value is multiplied by itself or squared. The spacing of nonlinear scales can be easily be spotted since they are uneven.

Take a look at a multimeter scale (see Fig. 6-7). A multimeter usually has a number of scales. Each scale on a multimeter is defined individually.

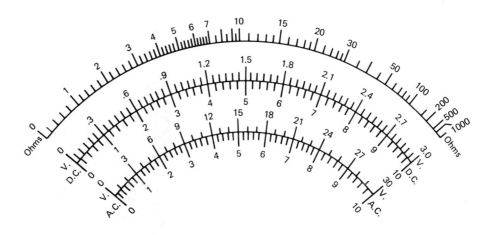

FIGURE 6-7 Multimeter scale.

PARALLAX ERROR

A parallax error results from the fact that there is a distance between the pointer and the scale behind it. If you look at the pointer from an angle, you will get an incorrect reading. From the side, the reading will be different than from directly in front. It is also possible to have a parallax error if you use both eyes. The best practice is to use one eye. Make sure your view is from directly in front of the pointer. Some meters have a small mirror attached to the scale to reduce parallax error (see Fig. 6-8). You minimize parallax error when you line up the pointer with the reflection in the mirror.

AMMETER

The ammeter is used to measure current. It measures current in microamperes (μA), milliamperes (mA), or amperes. The microampere is very small (0.000001 ampere). The milliampere is also very small (0.001 ampere). In most instances you will be working with milliamperes and amperes. Microamperes are usually found in circuits that use transistors and other semiconductor devices.

It is important to fit the meter to the job. You select a meter to fit the work you are doing and the range of currents you will be encountering on the job. Keep in mind that the 1-milliampere meter is used for measuring direct current (dc) of up to 1 milliampere. Such a meter does not have sufficient sensitivity to measure currents of less than 100 microamperes or 0.1 milliampere.

FIGURE 6-8 Meter with mirrored scale.

Extending the Range

The 1-milliampere meter movement can be used to measure direct currents that are larger than 1 milliampere. To do this, a resistor is connected in parallel with the meter movement. This resistor is called a *shunt*. To measure 2 milliamperes, the shunt requires a resistor that can pass 1 milliampere around the meter movement. Thus, 1 milliampere goes through the meter and 1 milliampere through the shunt (see Fig. 6-9).

Various-sized shunts can be connected, by switching, across the meter movement to increase its range from 1 milliampere to 1 ampere, 10 amperes, or even 100 amperes (see Fig. 6-10).

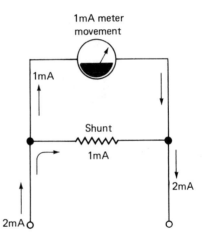

FIGURE 6-9 Shunting current around the meter movement.

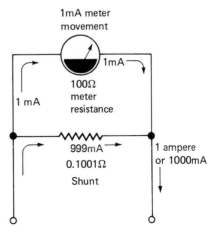

FIGURE 6-10 One-milliampere movement that can read up to 1 ampere.

114 ELECTRICAL MEASURING INSTRUMENTS

Connecting an Ammeter in the Circuit

An ammeter is always inserted in a circuit in series with the load.

Clamp-on Ammeter

The clamp-on ammeter is shown in Fig. 6-11. It is inserted over a wire carrying alternating current (ac). The magnetic field around the wire induces a small amount

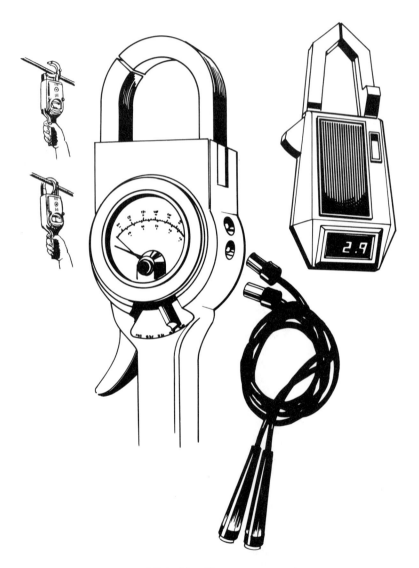

FIGURE 6-11 Clamp-on ammeter.

of current in the meter. The scale is calibrated to read amperes or volts. Because the wire is run through the large loop extending past the meter movement, it is possible to read ac or current, without removing the insulation from the wire or opening the circuit to install the meter. This meter is very useful when working with ac motors and compressors. Leads are provided so that the meter can be used as a voltmeter by connecting the leads across whatever voltage source is being checked.

VOLTMETER

The voltmeter measures electrical pressure, or volts. It is nothing more than an ammeter with a resistor added to the meter circuit. The high resistance of the voltmeter makes it possible to place it across a power source (in parallel) (see Fig. 6-12). A number of resistors, called multipliers or scaling resistors, can be switched into a meter circuit to increase its range to make it capable of measuring higher voltages.

A voltmeter circuit is shown in Fig. 6-13. It has 100-volt full-scale deflection. The resistor indicated in the illustration is the meter multiplier. The multiplier's resistance rating is 99.9 kilohms (kΩ). The notation k represents 1000. So the resistance is 99,900 ohms. Meter multipliers are usually precision-type resistors with a $\pm 1\%$ accuracy.

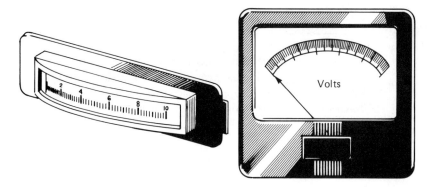

FIGURE 6-12 Voltmeter.

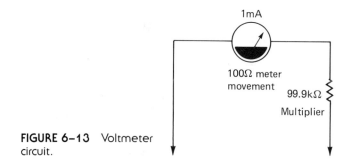

FIGURE 6-13 Voltmeter circuit.

Connecting a Voltmeter

A multirange voltmeter uses a switch to select among multipliers. Scales are arranged so that voltage is measured near the center of the meter scale. The center portion of a meter scale is always the most accurate part of the deflection range.

AC AMMETER

DC meters can be modified to measure ac current. This is done by connecting a rectifier in series with the meter movement (see Fig. 6-14). A rectifier is a device that changes ac to dc. An ac scale and rectifier are shown in Fig. 6-14. Today, most rectifiers are semiconductor diodes. In the past, vacuum tubes were often used for this purpose.

AC VOLTMETER

The ac voltmeter is nothing more than the dc meter movement with a diode inserted in series with the meter movement and the multiplier resistor (see Fig. 6-15).

OHMMETERS

It is sometimes necessary to troubleshoot a piece of equipment when no electricity is available. Also, in some cases, the application of power to the circuit may cause some damage. Therefore, you need to be able to troubleshoot the circuit without having to turn on the power and maybe burn up something. This is when an ohmmeter is used to check a circuit because it has its own power source (batteries).

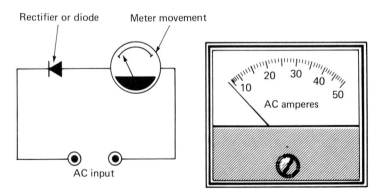

FIGURE 6-14 AC ammeter circuit.

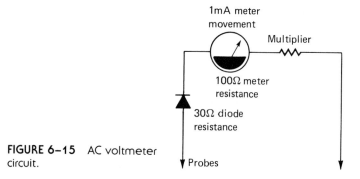

FIGURE 6-15 AC voltmeter circuit.

Two types of ohmmeters are made for use in the field: the *series ohmmeter* for measuring high resistances and the *shunt ohmmeter* for measuring low resistances like motor windings and checking for continuity of wiring.

The ohmmeter is most convenient when it comes to measuring resistance in a circuit. It has two probes that can be placed across an unknown resistance. The value of the resistance is displayed on the meter scale (see Fig. 6-16).

The ohmmeter has its own voltage source or batteries. It is *never* connected to a live circuit (one with the power on). Full-scale deflection on the series ohmmeter indicates 0. This means that when the probes are shorted the meter pointer goes to the right and indicates no resistance or zero. When the probes are apart with nothing in between, they indicate infinity (∞), which means too much resistance to measure. Note how the scale is nonlinear. The series ohmmeter measures high resistances and the meter scale reads from right to left.

The shunt ohmmeter is designed to measure low resistance, usually below 200 ohms. See the scale in Fig. 6-17. Note, too, that the scale reads from left to right, with the zero being on the left and infinity on the right. An imporatant thing to remember about the shunt ohmmeter is to be sure you turn it *off* when you have finished using it. It will run down the batteries if left in the *on* position. The shunt

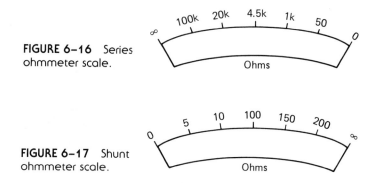

FIGURE 6-16 Series ohmmeter scale.

FIGURE 6-17 Shunt ohmmeter scale.

ohmmeter can be identified because it requires an on-off switch. If it is part of a multimeter, then switching it to amperes or volts will turn it off.

MULTIMETER

Voltmeters, ammeters, and ohmmeters are often combined into a single instrument. Such an instrument is called a multimeter (see Fig. 6-18). Multimeters have a range switch that selects the scale. Another control is a function switch to control its use as an ohmmeter, ammeter, or voltmeter. Multimeters also have a *zero adjust knob* for the ohmmeter circuits. There are also jacks for the probes. The functions and ranges of a multimeter are determined by selector switches.

MEGGER

The megger (see Fig. 6-19) is a device that is used for very high resistances. It is used to check for insulation breakdown in motors and compressors. A handcrank is attached to a coil inside the megger. The coil is placed inside a permanent magnet field. A high voltage is produced when the crank is rotated. This high voltage is required to drive small amounts of current through extremely high resistances. The meter movement has to be very sensitive.

The megger can produce a shock if the leads are touched while the handle is turned. Be careful when using a megger. Do not attach the leads to anything that will be easily damaged by high voltage.

FIGURE 6-18 Multimeter.

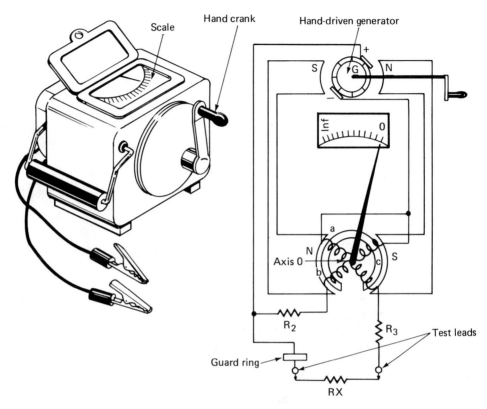

FIGURE 6-19 Megger.

DIGITAL METER

Figure 6-20 shows a digital meter. There are no coils or magnets in digital meters. The meter presents its reading in numbers on a liquid crystal display. This is the type of digital readout found on many instruments, clocks, and watches.

This type of meter can be used to measure ac and dc voltages and currents with great accuracy. In fact, most digital meters can measure voltages to one-hundredth of a volt. They can also measure resistances within one-tenth of an ohm. They are sometimes more accurate than needed for the job. They are also presently more expensive than the analog type. This, of course, is subject to change, as is the case with all electronics.

OTHER INSTRUMENTS

If you are going to be working with heating, air-conditioning, and refrigeration circuits, you will need some other types of instruments to do the job properly and,

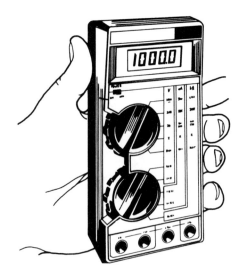

FIGURE 6-20 Digital meter.

most of all, efficiently. Electronics is being used today to make the job of recharging refrigeration systems easier and more accurate. A few of these newer electronic devices now available to refrigeration and air-conditioning technicians are described next.

Automatic Halogen Leak Detector

A number of halogen leak detectors are available. However, some of the newer electronic detectors are very sensitive. Figure 6-21 shows an automatic type that has a permanently sealed, miniature battery-operated pump that produces a computer-like beeping signal that changes in both speed and frequency as the leak source (halogen or vapor) is approached. It can detect leaks as small as 0.5 ounce per year. An additional feature is that this instrument is capable of calibrating itself automatically while in use. It operates on two C cells. The detector sensor is not ruined by large doses of refrigerant.

Electronic Sight Glass

The "electronic" sight glass emits an audible signal indicating that the refrigeration system being charged is full (see Fig. 6-22). It uses sonar to "look" inside refrigeration tubing and, in effect, can be considered an "electronic" sight glass. It is very handy for capillary systems, which have no sight glass. It is also helpful when charging household and commercial refrigerators and freezers. It is also valuable when checking automobile systems and window air conditioners. This instrument detects flash gas and bubbles in the liquid line. It detects droplets in the suction line and starved evaporators. It can also be used to check for refrigerant floodback and becomes very useful for multiple evaporator balancing and for adjusting thermostatic expansion valves.

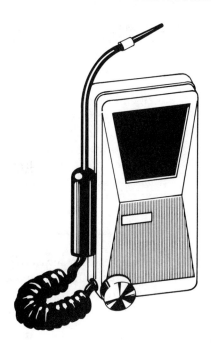

FIGURE 6-21 Electronic halogen leak detector.

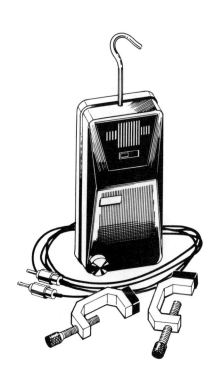

FIGURE 6-22 Electronic sight glass.

Electronic Charging Meter

The electronic charging meter is a new tool for dispensing refrigerant into a system in fractions of an ounce without limit from a standard 30-pound cylinder. The charging meter measures the charge dispensed into the system by weight, but it is totally different from a bathroom-type scale. It reads directly like a service station pump. No more wasted time refilling a charged cylinder; no more recorrecting for temperature or type of refrigerant. The refrigerant charge appears on a liquid crystal display, which is easily read even in bright sunlight. There are no controls or adjustments. It is operated at the touch of a button. It can also be used as a conventional weight scale so that your cylinder can be weighed to ascertain the amount of refrigerant remaining for other jobs. It operates on six C cells (see Fig. 6-23).

USING AN OHMMETER

The basic unit for measuring resistance is the ohm. An ohmmeter is a device used to measure resistance or ohms. It is an ammeter (or milliammeter or microammeter) movement, modified to measure resistance (see Fig. 6-2). The ohmmeter is easily damaged if not properly connected to the circuit under test. Most multimeters are capable of measuring ohms with at least three different ranges: $R \times 1$, $R \times 10$, and $R \times 10,000$. Thus, it is capable of measuring up to 20 million ohms when on the $R \times 10,000$ scale. The meter scale has to be multiplied by the 100 or 10,000 number to obtain the proper value. The $R \times 1$ scale is used to check for shorts and very low resistances, such as in compressor motors, relay contacts, and motor windings and switches.

WARNING

An ohmmeter has a battery in it, so it does not use power from the circuit under test. In fact, do not connect *it to a* live *circuit (one with the power on). To do so will result in* destruction *of the meter movement.*

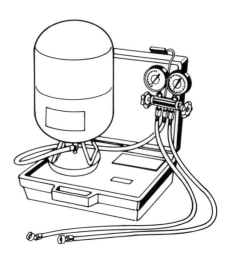

FIGURE 6-23 Electronic charging meter.

Adjusting the Meter

Adjust the ohmmeter so that the meter reads *zero* before starting to use it to measure resistance. This means that you have to adjust the meter circuit to compensate for the battery voltage changes. Battery voltages decrease with shelf life. It doesn't matter whether or not the battery is used. It will, in time, lose its voltage.

The ohmmeter can be used to test compressor windings for opens or shorts. An open means the meter does not move or reads infinite. A short means the meter reads zero. If you read the same resistance between the three pins of the compressor, it means the motor is three phase and will not run on single-phase power.

One important thing is to obtain an ohmmeter that can measure very low resistances, such as 1 to 10 ohms, with some degree of accuracy. This calls for a digital meter that is somewhat expensive or a multimeter with a shunt ohmmeter capability. The shunt ohmmeter has a scale for low resistance that reads from left to right (see Fig. 6-17). Since, in some instances, a 1-ohm reading can indicate a short, you will probably need a good digital ohmmeter to be sure. The common readings for a compressor are 1, 5, 10, 15, and 20 ohms resistance.

Check Fig. 6-24 for the layout of compressor terminals. Note that the resistance from R to S is 11 ohms. This means that both windings (start and run) are in

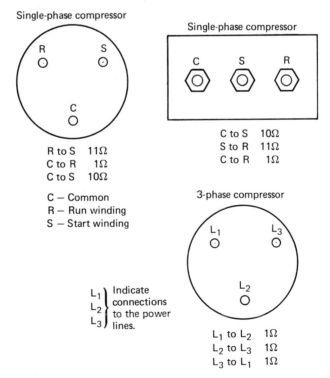

FIGURE 6-24 Reistances between compressor terminals.

series and are being added at this point. The reading from R to C is 1 ohm. This means that the run (R) winding has 1 ohm of resistance and is normal. If the reading is less than 1 ohm, it means you have a short in the start winding. The reading from C (common) to S (start winding terminal) is normally 10 ohms. If you obtain *infinite* for any of these readings, it means the winding is *open*. If you read less than 3 ohms for the run winding, it means the winding is *shorted*. If you get a reading from any one of the three terminals to the case of the compressor, it means the winding is shorted to the case.

Other steps in troubleshooting will be explained in following chapters.

REVIEW QUESTIONS

1. How is electricity measured?
2. What is a d'Arsonval movement?
3. What is meant by full-scale deflection?
4. What is parallax error?
5. What is the difference between milliampere, microampere, and ampere?
6. How is the range of a meter extended?
7. How is the ammeter connected in a circuit?
8. What is the difference between a series and shunt ohmmeter?
9. What is a megger? Where is it used?
10. How is the voltmeter used to test a circuit?

PERFORMANCE OBJECTIVES

Know the type of current furnished by a battery.
Know the difference between a cell and a battery.
Know what is meant by the ampere-hour rating of a battery.
Know how cells are connected in series and in parallel.
Know what causes batteries to have a shortened life.
Know what a *sine wave* is.
Know the difference between *alternating current* and *direct current*.
Know what *rms* means.
Know the meaning of *polyphase*.
Know the difference between *delta* and *wye* connections.
Know how three-phase power is generated.

CHAPTER 7

Electrical Power: Direct Current and Alternating Current

Today's electronics require sources of direct current to operate the semiconductor devices associated with their control circuits. Direct current can be generated with any number of battery types. However, only the more familiar types are presented here. The heating, air-conditioning, and refrigeration technician must learn to choose the right battery for the job.

TYPES OF BATTERIES

A single cell produces a voltage between 1.2 and 2.2 volts (see Fig. 7-1). Two or more cells connected either in series or parallel make a battery. The number of cells in a battery is related to its voltage and current requirements. Cells connected in series produce higher voltages. Cells connected in parallel produce higher current.

Primary Cell

A cell is a device that converts chemical energy into electrical energy. The primary cell cannot be recharged. Once the primary cell is discharged, it is thrown away. The chemical action that discharged the primary cell cannot be reversed (see Fig. 7-2).

Secondary Cell

The secondary cell can be recharged. The chemical action is reversible. The electrodes and electrolyte that make up the cell can be restored to the same makeup that existed before the discharge (see Fig. 7-3).

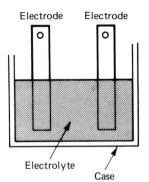

FIGURE 7-1 Makeup of a cell.

The secondary cell, such as the one used in your automobile battery, can be discharged and charged many times. However, the secondary cells that make up such a battery deteriorate. The number of times a secondary cell can be charged and discharged depends on its design, construction, use, and how well it is maintained.

During discharge, electrons flow from the − (negative) terminal of the cell or battery to the + (positive) terminal by way of a load (see Fig. 7-4). When the cells are charged, the process is reversed and electrons flow from + to −. This reversal of electron flow causes a chemical action to take place that produces a charged condition when completed.

DRY CELLS

The dry cell is a *primary* cell. It is so named because the electrolyte is not a liquid but a dry, pastelike compound. The many types of dry cells differ mainly in the types of electrodes and electrolyte used. Three types of dry cells in popular use are the carbon–zinc cell, the alkaline cell, and the mercury cell.

The alkaline cell will last longer than the carbon–zinc because it has a lower internal resistance and higher current output. This is due to the potassium hydroxide used as an electrolyte. Potassium hydroxide has a higher conductivity than the sal ammoniac used in the carbon–zinc cell.

The mercury cell has a positive electrode of mercuric oxide. The negative electrode is made of amalgamated zinc. The entire assembly is mounted in a sealed steel container. It is produced in flat and cylindrical shapes. With the use of pressed powder electrodes, the mercury cell can be made very small. This is an advantage for use in hearing aids and other miniature electronic devices, such as watches and timers used with electronic *temperature controls* (see Fig. 7-5).

BATTERY SPECIFICATIONS

Batteries are made in many sizes and shapes. Figure 7-6 shows the standard cell sizes. Batteries come in even larger sizes. They may be as small as the 9-volt transistor battery and as large as a submarine battery that measures over 6 feet high.

BATTERY SPECIFICATIONS **129**

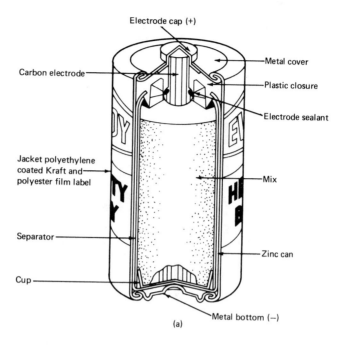

(a)

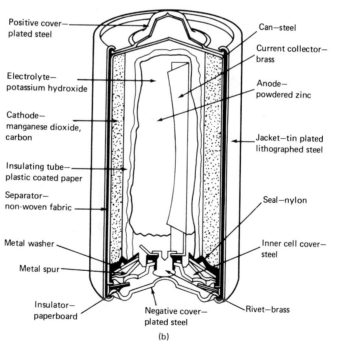

(b)

FIGURE 7-2 Dry cells.

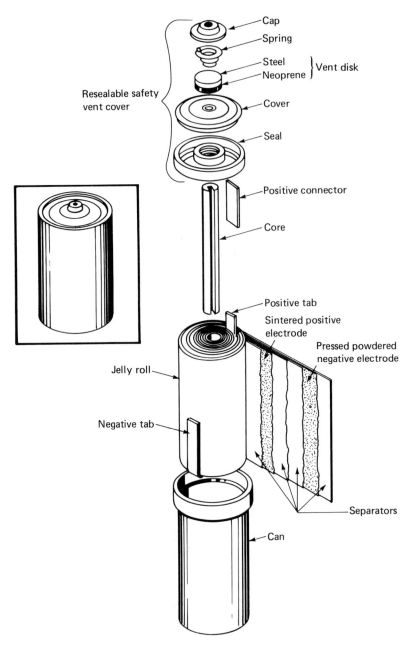

FIGURE 7-3 Rechargeable nickel–cadmium cell. (Courtesy of Eveready)

FIGURE 7-4 Cells connected − to + with the load.

Cell Top (Negative Terminal) Single type. Steel coated with copper on inside and with nickel and gold externally.

Anode Powdered zinc (amalgamated), together with gelled electrolyte.

Nylon Grommet Coated with sealant to ensure freedom from leakage. Color code: Mercury, blue (high rate) or yellow (low rate); Silver, green (high rate) or clear (low rate).

Sleeve Nickel-coated steel. Supports grommet pressure. Also aids in consolidating cathode.

Absorbent Separator Felted fabric (cotton or synthetic). Prevents direct contact between anode and cathode. Holds electrolyte.

Barrier Separator Membrane permeable to electrolyte but not to dissolved cathode components.

Electrolyte Alkaline solution. In anode, cathode and separators.

Cathode Mercuric oxide with graphite. Highly compacted.

Cell Can (Positive Terminal) Nickel, or steel coated on both sides with nickel.

FIGURE 7-5 Mercury cell.

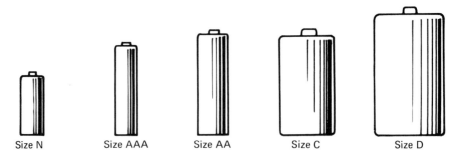

FIGURE 7-6   Five standard cell sizes.

Batteries are rated according to voltage. They are also rated according to the recommended current range or in terms of *ampere-hour* capacity.

Higher current demands and longer operating hours than the specifications call for will shorten the life of the battery. Higher current and longer operating hours will reduce the ampere-hour capacity of the battery. The ampere-hour rating is good for a reasonable range on each side of the point used to establish the rating.

Alkaline batteries are made for continuous-drain and high-current operation. Their service life may be 10 times greater than a comparable carbon–zinc battery. For light loads, the service life may be only a limited number of times longer. For light loads, then, alkaline batteries may not be practical because of the higher initial costs.

The mercury battery is expensive in comparison to the alkaline and carbon–zinc.

CONNECTING CELLS

Cells can be connected in series or parallel to obtain a desired voltage or current rating. When cells are connected in series, the individual voltages add. Four 1.5-volt cells connected in series provide 6 volts. The 9-volt transistor battery is made up of six 1.5-volt cells connected in series.

Cells are connected in series by putting the − terminal to the + terminal of the next cell. They are connected, − to +, − to +, and − to +, for as many cells as you need to produce the voltage needed (see Fig. 7-7).

Cells connected in parallel, as in Fig. 7-8, provide more current. They are connected by placing the − to − terminals together and the + to + terminals together. You obtain the same voltage as from one cell, but the current available from all the cells is added. Do not connect cells with different voltages in parallel. They have a tendency to drain when not in use.

Higher voltage and current batteries such as lantern batteries can be obtained by using cells connected in series–parallel (see Fig. 7-9).

CONNECTING CELLS 133

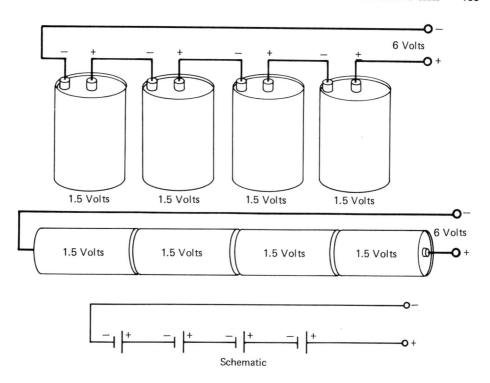

FIGURE 7-7 Cells in series.

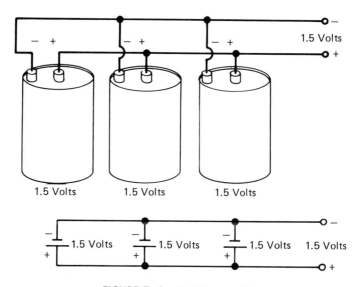

FIGURE 7-8 Cells in parallel.

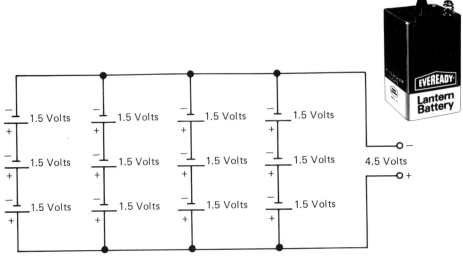

FIGURE 7-9 Cells in series–parallel.

BATTERY MAINTENANCE

Batteries that have been properly treated will last longer. Two of the most important elements working against batteries are overheating and *sulfation*. The formation of lead sulfate on the battery plates is normal during the discharge cycle of the lead–acid battery. The lead–acid type is used in the automobile and elsewhere when high currents are needed for an extended period of time. If a battery is allowed to stay in a discharged state for a period of time, the deposit hardens and no longer responds to recharging. The battery is said to be sulfated and cannot be restored to full capacity. Sulfation extended over a long period makes the battery useless.

Other practices that can cause shortened battery life are frequent *undercharging*, too high a specific gravity, long idle periods in a discharged condition, and defects caused by rapid cycles of high-current discharge and high-current charge.

Charge and discharge currents should not be allowed to overheat the battery. Overheating can cause bending of the battery plates and excessive sulfation. Avoid rapid discharge and keep the proper water level. There is a normal loss of water from the electrolyte, and it is important that the water be replaced. Reasonably pure water should be added when necessary to maintain the proper level and the correct specific gravity range. Overheating tends to encourage faster water evaporation. Whenever the battery is on charge, there is some release of hydrogen and oxygen gas at the electrodes. These gases bubble through the electrolyte to the surface and then through the vent hole in the cell. Indirectly, this also means a loss of water, which must be replaced. However, recent battery technology has improved the battery to such a point that maintenance-free operation is possible for the life of the battery if it is cared for and used for its intended purpose. The sealed battery does not call for the addition of water inasmuch as there is no evaporation.

NICKEL-CADMIUM CELLS

The nickel-cadmium cell comes in a variety of sizes. It can be sealed and made very compact. Figure 7-3 shows the nickel-cadmium construction. It is a high-efficiency cell made so that it can deliver high current at a constant-voltage output. It can be recharged often and still have a long active life. Storage life is not critical and long idle periods, charged or discharged, have no bad effects. It can also be completely discharged without damage.

The voltage output of the nickel-cadmium cell is usually 1.25 volts or a little less than the 1.5 volts of the carbon-zinc cell.

ALKALINE CELL

The alkaline cell can be either a primary or secondary cell. It has characteristics comparable to a similar sized nickel-cadmium battery. The number of times that the alkaline battery can be charged and discharged is less. However, the initial cost is much less than a similar sized nickel-cadmium. In some cases, overall economic considerations favor the alkaline type of battery.

The alkaline rechargeable cell should never be discharged below 0.9 volt per cell because the battery life will be shortened. If an alkaline rechargeable cell is completely discharged, it is impossible to recharge it.

When an alkaline secondary battery is discharged for the maximum recommended time and then recharged for the maximum recommended time, it may be cycled as many as 50 times before the battery falls below the rated cutoff voltage at the end of the recommended discharge period. The alkaline cell output voltage is 1.5.

Proper battery care is needed to make sure that electronic and electrical equipment operate properly and when needed.

ALTERNATING CURRENT

Direct current has electricity flowing in one direction. Another type of electricity that you must understand to be able to troubleshoot equipment is alternating current (ac). Most motors and other control devices on refrigerators, air conditioners, and heating units operate on alternating current, or ac as it is most commonly called.

Alternating current is constantly changing in direction and amplitude. Inasmuch as dc was discovered first, the formulas and mathematical equations for working with electricity were developed for dc. However, after the introduction of ac, around the beginning of 1900, a number of changes had to be made in the way electricity was handled mathematically. Alternating current was equated to dc. This was for the sake of making previous formulas work and, at the same time, to produce an understanding of what ac could do when applied to motors and other devices.

SINE WAVE

The shape of the ac waveform is that of a *sine wave*. There are many other types of waveforms, such as the pulse, sawtooth, and square. However, these other waveforms are considered combinations of sine waves of different amplitude, frequency, and phase. Thus, the forms other than sine waves are usually discussed in terms of their sine-wave (sinusoidal) makeup. The basic sine-wave current is a single frequency, and it is this sine wave of single frequency (60 Hz) that we are interested in for our use (see Fig. 7-10).

The sine wave is unique. It can take many shapes. The sine wave has a very special relation to time and motion. Take a close look at Fig. 7-11. Assume that a point at the end of the radial (r) takes one second to travel at a constant speed around the circle, as shown in Fig. 7-12A. How long would it take the point to move 90° or 180°? Since the radial rotates at a constant speed, it would require 0.25 second to rotate 90° (from a horizontal position to a vertical position). At this instant, the point at the end of the radial (on the circumference of the circle) would be at the highest point with relation to the horizontal line. At 180° the point will have been in motion for 0.5 second and will have reached the horizontal line. At 270°, or 0.75 second, the point at the end of the radial line will be at the lowest point with relation to the horizontal reference line. Finally, in 1 second the point at the end of the radial comes back to its starting position, and one cycle (called 1 hertz) has been completed. Keep in mind that time can be stated in degrees of rotation and the horizontal reference line is time.

Electricity available for home use in the United States is 60 hertz. Thus, the generator must turn 60 times per second to produce 60 hertz. One hertz is one cycle per second. Thus, one sine wave is produced every sixtieth of a second.

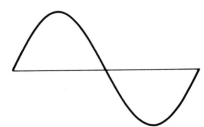

FIGURE 7-10 Sine wave.

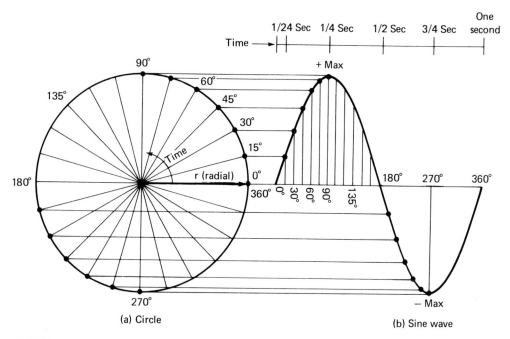

FIGURE 7-11 Sine wave showing how it is generated by a point rotating in a circle, that point being one coil on a generator rotating in a magnetic field.

SINE-WAVE CHARACTERISTICS

Figure 7-12 shows the output of two cycles of the generator. These two cycles (hertz) of alternating current have a negative alternation and a positive alternation. Each sine wave has a positive alternation that swings above the zero axis and a negative alternation that swings below the zero axis. Unless otherwise marked, it is normal

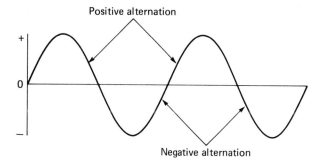

FIGURE 7-12 Two hertz (cycles) of a sine wave.

for the alternation that swings above the zero axis to be positive, and the alternation that swings below the zero axis to be negative. Thus, a complete cycle (hertz) is made up of a positive and negative swing.

Two important characteristics of a sine wave are period and frequency. It takes time for the sine wave to go through its positive and negative alternations and return to the origin or zero reference line. The time required for a single cycle (hertz) is called the *period* of the sine wave. The number of complete cycles (hertz) made each second is called the *frequency* of the sine wave.

Phase

In electrical work it is often necessary to relate one sine wave to another. Two sine waves are shown in Fig. 7-13. This relation between two sine waves, known as *phase*, can be a comparison between a sine-wave voltage and a sine-wave current. Or it can be a comparison between two sine-wave voltages or between two sine wave currents. This phase or difference in time is another important characteristic of a sine wave.

Maximum and Peak Values of AC

Three values are used to describe ac:

1. Peak value
2. Average value
3. Root-mean-square (rms) value

Peak Value. The maximum point on a sine wave is its peak value. Both peaks of a single hertz may be included in a reference. If so, it becomes a peak-to-peak (p-p) value. A peak value of 100 volts means that the peak-to-peak value is 200 volts. This is shown in Fig. 7-14.

Average Value. The average of all instantaneous values of a generator is measured at regular intervals (see Fig. 7-11). Instantaneous values are taken at selected points in the generating process. The average of these is the average value of ac current or voltage.

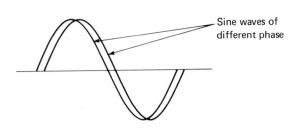

FIGURE 7-13 Phase difference between two sine waves.

Sine waves of different phase

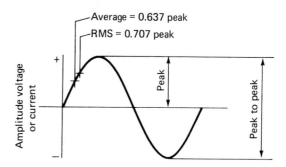

FIGURE 7-14 Sine-wave values of average, rms, and peak.

The half-hertz is used to compute this value. This is because the average value for full-hertz (cycles) would be zero. The zero average would come from the fact that half the values are positive and half are negative.

To compute the average value, sine-wave values for angles between 0° and 180° are added. The total is then divided by the number of values taken. As seen in Fig. 7-14, the average is equal to 0.637 of the peak. For example, suppose the peak voltage of a sine wave is 100 volts. The average voltage is 63.7 volts (100 × 0.637 = 63.7).

Root-Mean-Square (RMS) Value. A value equivalent to effective voltage or effective current is known as the root-mean-square (rms) value. It is calculated mathematically. Imagine one-quarter of the sine wave broken down into 90 parts. This is one part for each degree of a hertz. The value of each degree is squared, or multiplied by itself. All the squared values are added together. Then the total is divided by 90. This produces the average of the squares. A square root is then figured for this average. This is the rms value.

POLYPHASE ALTERNATING CURRENT

The armature coils of a single-phase alternator (ac generator) are connected in series. At any given instant the voltage induced is the sum of the voltages induced separately in each coil. If the coils were not connected in series, and if their connections were brought out separately, the voltages induced would vary in phase with relation to each other. This is because each coil is in a different position with respect to the flux lines of the field (see Fig. 7-15).

When windings, identical in size and of the proper phase relationships, are added to the single-phase alternator, a voltage output is obtained from each coil added. These voltages differ in phase but not in amplitude. The output from such a generator is known as a *polyphase*-output system. Systems of this type are generally two- or three-phase systems with the three-phase the most widely used. In fact,

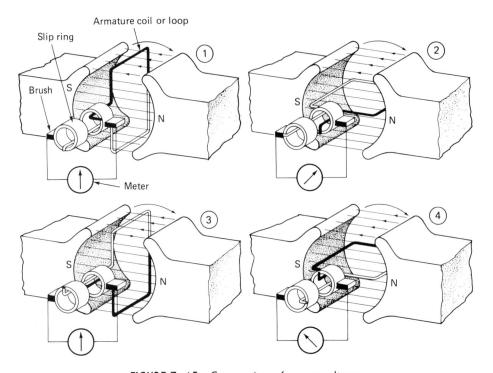

FIGURE 7-15 Generation of an ac voltage.

all the power generated by the large utilities is three-phase. Large air-conditioning and refrigeration installations use three-phase power.

Three Phase

A three-phase system is one in which the voltages have equal magnitudes and are displaced 120 electrical degrees from each other (see Fig. 7-16). The three windings are placed on the armature 120° apart (Fig. 7-17). As the armature is rotated, the outputs of the three windings are equal but out of phase by 120° (see Fig. 7-18).

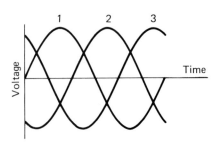

FIGURE 7-16 Three-phase voltages as induced in a generator stator windings.

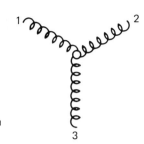

FIGURE 7-17 Schematic representation of a three-phase stator connected in a wye (Y) configuration.

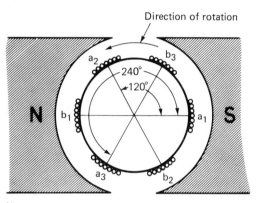

FIGURE 7-18 Basic three-phase alternator.

Note: Windings a_1b_1, a_2b_2, and a_3b_3 are individual windings.

THREE-PHASE CONNECTIONS

Three-phase (3ϕ) windings are usually connected in either a delta or wye configuration. Each of these connections has definite electrical characteristics from which the designations *delta* and *wye* are derived (see Figs. 7-19 and 7-20).

ELECTRICAL PROPERTIES OF DELTA AND WYE

Delta Connection (Δ)

In a balanced circuit, when the generators are connected in delta, the voltage between any two lines is equal to that of a single phase. The line voltage and the voltage across any winding are in phase, but the line current is 30° or 150° out of phase with the current in any of the other windings (see Fig. 7-21). In the delta-connected generator, the line current from any one of the windings is found by multiplying the phase current by the square root of 3, which is 1.73.

142 ELECTRICAL POWER: DIRECT CURRENT AND ALTERNATING CURRENT

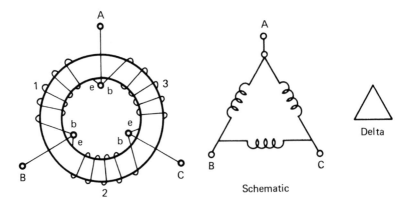

FIGURE 7-19 Delta-connected ring.

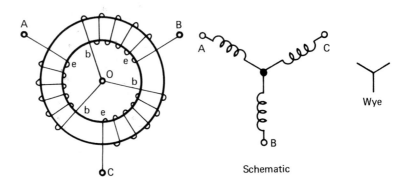

FIGURE 7-20 Wye-connected ring.

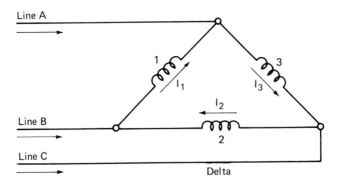

FIGURE 7-21 Delta-connected currents.

ELECTRICAL PROPERTIES OF DELTA AND WYE

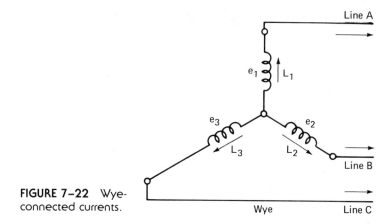

FIGURE 7-22 Wye-connected currents.

Wye Connection (Y)

In the wye connection, the current in the line is in phase with the current in the winding. The voltage between any two lines is not equal to the voltage of a single phase, but is equal to the vector sum of the two windings between the lines. The current in line A of Fig. 7-22 is the current flowing through the winding L_1; that in line B is the current flowing through the winding L_2; and the current flowing in line C is that of the winding L_3. Therefore, the current in any line is in phase with the current in the winding that it feeds. Since the line voltage is the vector sum of the voltages across any two coils, the line voltage E_L and the voltage across the winding $E\phi$ are 30° out of phase. The line voltage may be found by multiplying the voltage of any winding $E\phi$ by 1.73.

Delta and Wye Summarized

The properties of delta connections may be summarized as follows: The three windings of the delta connection form a *closed loop*. The sum of the three equal voltages, which are 120° out of phase, is zero. Thus, the circulating current in the closed loop formed by the windings is zero. The magnitude of any line current is equal to the square root of 3 (1.73) times the magnitude of any phase current.

Properties of the wye connection do not form a closed loop. The magnitude of the voltage between any two lines equals the magnitude of any phase voltage, times the square root of 3, or

$$E_L = \sqrt{3} \times E\phi$$

The current in any winding equals the current in the line.

Applications of three-phase power to motors will be discussed in more detail in Chapter 12. Connections of power supply transformers to furnish three-phase power are discussed in Chapter 8. As previously mentioned, three-phase power is used in commercial and industrial air-conditioning and refrigeration equipment and in some heating units.

REVIEW QUESTIONS

1. What type of current is obtained from a battery?
2. What is the difference between a cell and a battery?
3. What are three types of dry cells in popular demand today?
4. What is meant by the ampere-hour rating of a battery?
5. How are cells connected in series?
6. Why are cells connected in series?
7. Why are cells connected in parallel?
8. What causes batteries to have a shortened life?
9. What is a sine wave?
10. What is the difference between alternating current and direct current?
11. What is a hertz? Where do you see the term used most frequently?
12. What are the three values used to describe ac?
13. What does rms mean? Where is the term used?
14. What is meant by polyphase ac?
15. What is the difference between a wye connection and a delta connection?
16. Where will you find three-phase power used?

PERFORMANCE OBJECTIVES

Know the two names for *coil*.
Know what the term *inductance* means.
Know the four things that determine inductance.
Know what *self-inductance* is.
Know what *mutual inductance* is.
Know the effects of a counter emf.
Know what *out-of-phase* means.
Know what *inductive reactance* is.
Know what a transformer does.
Know what a *variac* is.
Know the three losses associated with a transformer.
Know what is meant by *inductive kickback*.

CHAPTER 8

Inductors and Transformers

The magnetic field that surrounds a coil produces an opposition to a change in current in that coil. This *opposition* is not the same as *resistance* is to direct current. A coil has an opposition to alternating current because the ac is constantly changing its direction and amplitude. The opposition that a coil presents to ac is called *inductive reactance*. Inductive reactance must be taken into consideration when a motor or transformer is in an electrical circuit. It is also very important in the design of start circuits for compressors. Coils are used for filter circuits in electronic power supplies. There are many uses for inductors and for inductive reactance.

INDUCTORS

Coils are also referred to as inductors and chokes. The inductor is used in a circuit to retard the flow of ac and to allow dc to pass. Inductors are diagrammed in circuits through the use of the symbol shown in Fig. 8-1.

The property of a coil that opposes any change in circuit current is known as *inductance*. The symbol for inductance is L. Inductance is measured in *henrys*, abbreviated H. Since the henry may be too big for most electronic circuits, it is

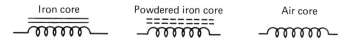

FIGURE 8-1 Symbols for inductors.

broken down further into millihenrys (mH) and microhenrys (μH). Motors and transformers have inductance and therefore inductive reactance when ac is applied.

CHANGING INDUCTANCE

The inductance of a coil depends on four factors:

1. *Number of turns:* The more turns a coil has the greater its inductance.
2. *Diameter of the coil:* The wider the cross section of a coil, the higher the inductance is.
3. *Permeability of the core:* The more permeable the core the better are the magnetic properties and the higher the inductance.
4. *Length of the coil:* The shorter the coil the higher the inductance is. Coil length and inductance are said to be inversely proportional.

Figure 8-2 shows how the inductance of a coil can be increased by adding turns, by adding a core, and by the way layers are wound.

SELF-INDUCTANCE

The ability of a conductor to induce voltage in itself when the current changes is called self-inductance. This ability can be valuable. Self-inductance is also measured in henrys.

Self-inductance is produced by varying current in a coil. The current produces a magnetic field around each turn of the coil. The field around each turn then cuts across other turns (see Fig. 8-3). A changing current in the turns of a coil produces a changing magnetic field. Self-induced voltage occurs when the changing magnetic field cuts the turns of the coil and induces a voltage across the coil.

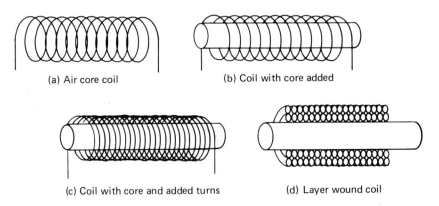

(a) Air core coil
(b) Coil with core added
(c) Coil with core and added turns
(d) Layer wound coil

FIGURE 8-2 Changing the inductance of a coil.

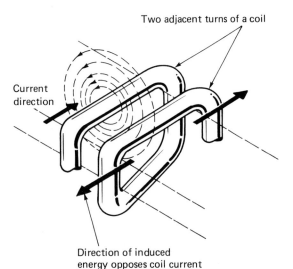

FIGURE 8-3 Self-inductance created by placing coils adjacent to one another. Note how the magnetic field cuts adjacent turns.

The amount of induced voltage depends on the rate of change of current. The faster the current changes, per unit of time, the higher is the self-induced voltage. Take a look at Fig. 8-4 and see how two points on a sine wave demonstrate the difference in ratio of current change.

The self-induced voltage is in the opposite direction to the magnetic field that produced it. This voltage is called a *counter* emf (cemf).

The cemf is shown in Fig. 8-5A. A current increases rapidly from zero to maximum. This causes the magnetic field to expand. The expanding magnetic field produces a cemf that moves away from the input current. The cemf cuts across

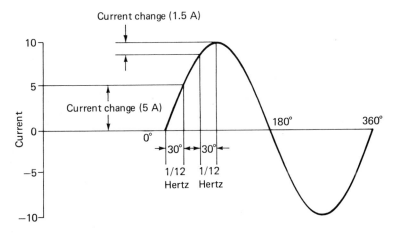

FIGURE 8-4 Sine-wave current changes.

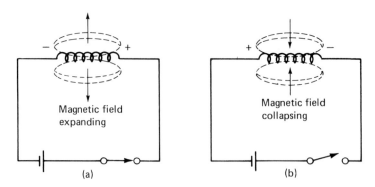

FIGURE 8-5 Counter emf.

windings in the path of the current. The increasing cemf opposes the increase in current in the circuit.

Now take a look at Fig. 8-5B. The circuit is broken by a switch. When this happens, the magnetic field collapses. At this time, the current in the circuit changes from its maximum value to zero. The collapsing field induces voltage across the coil. The induced voltage opposes the decrease in current. The current is thus prevented from dropping quickly to zero. The gradual decline in current level due to self-inductance is shown in Fig. 8-6.

In an inductor the lag created between current and voltage is 90°. Thus, in a purely inductive circuit the voltage will lead the voltage as shown in Fig. 8-7.

The coil opposes increases or decreases in current. This is valuable in filter circuits. In such a situation, inductance protects devices against sudden increases or drops in electrical current. An inductor, then, creates a lag or delay between the voltage and current. To show this, consider a purely inductive ac circuit, one with an inductor only. In such a circuit, current lags voltage by 90°. When this condition exists, the voltage and current are said to be *out of phase*.

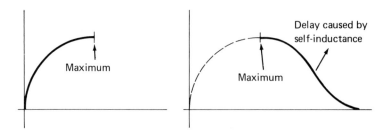

FIGURE 8-6 Time lag in an inductor.

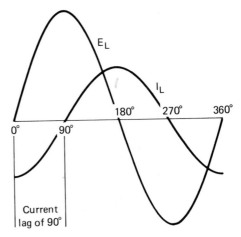

FIGURE 8-7 Voltage leads current by 90°.

MUTUAL INDUCTANCE

A condition in which two circuits share the energy of one circuit is called *mutual inductance*. The energy in one circuit is transferred to the other circuit. The two coils have a mutual inductance of 1 henry if a current change of 1 ampere per second in one coil induces 1 volt in the other coil. Magnetic lines of flux cause a coupling between the circuits.

A mutual inductance circuit is shown in Fig. 8-8. The symbol at the left of this diagram indicates an ac source. Coil L_2 is placed near coil L_1. The magnetic flux surrounds the wires making up L_2. This induces a voltage across coil L_2.

As soon as the magnetic field is built up by the ac, the current flow is reversed. The magnetic field begins to collapse. As the magnetic field collapses, it also induces a voltage in L_2. This induced voltage moves in the opposite direction from when it built up to a maximum.

L_2 is part of a circuit with an electrical load. Current flows through the load. Current in the circuit of L_2 also produces a magnetic field. This is in opposition to the field that induced the current in L_2. If a heavier load is placed in the circuit with

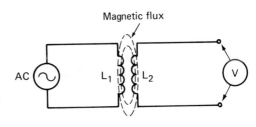

FIGURE 8-8 Mutual inductance.

L_2, a greater opposition is presented to the induced current. Thus, L_1 must draw more current for its source. It needs this additional current to make the increased opposition provided by the increased load across L_2. There is no load on L_1 until L_2 has a load connected across its terminals. This is the basis of how the transformer operates.

INDUCTIVE REACTANCE

As we have already seen, an inductor reacts to a changing magnetic field produced by a changing current. When alternating current passes through an inductor, a phase shift occurs. Voltage and current that were in phase become out of phase. Phase shifting occurs repeatedly. This results from the opposition to ac by an inductor. This opposition is referred to as *reactance*. X is the symbol for *reactance*. Since the opposition takes place in an inductor, it is called inductive reactance and is represented by X_L. It is an opposition so it is measured in ohms.

This type of opposition, or reactance, does not affect dc. The only opposition to dc comes from the resistance of the conductor (wire) that makes up the coil.

Measuring X_L

A number of factors determine X_L. One is the frequency of the alternating current. Another is the size of the inductor. The formula used to calculate X_L is

$$X_L = 2\pi f L$$

Inductive reactance is not measured by an ohmmeter, but is arrived at by mathematical calculation. In this equation, f is for frequency, measured in hertz. L is inductance, measured in henrys. π is a standard mathematical term with a value of 3.14. So 2π equals 6.28. Thus, if either frequency (f) or inductance (L) is increased, inductive reactance (X_L) also increases. If either of these factors is reduced, inductive reactance also becomes smaller.

POWER IN AN INDUCTIVE CIRCUIT

In an inductive circuit, power is absorbed by the coil during the time that current is increasing. This power is stored in the generated magnetic field. However, all the power absorbed during the current rise is *returned* to the source by the collapse of the magnetic field. This means that the average power consumed by the inductance is zero. Whatever it takes, it puts back.

The current, voltage, and power in a purely inductive circuit are shown in Fig. 8-9. During the first quarter-hertz of the sine wave (0° to 90°), the current is rising and power is being supplied from the source. The instantaneous power, which is

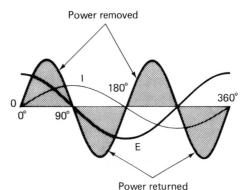

FIGURE 8-9 Power in an inductive circuit.

represented by the shaded areas in the illustration, is the product of the instantaneous voltage and instantaneous current.

The areas on the positive side of the zero axis represent power being transferred to a magnetic field from the electrical source. The areas on the negative side of the zero axis represent power being returned to the circuit. This occurs when the current decreases from its peak value toward zero on the 90° to 180° part of the hertz or cycle. The magnetic field then decreases in intensity and delivers power back to the source. There is now a transfer of magnetic energy back to electrical energy (the magnetic field is said to be collapsing). The source is now receiving power from the collapsing magnetic field. Since the power supplied and the power taken back are identical, the average power is zero.

Before you think an electric motor or fluorescent lamp can be operated without consuming energy, you have to bear in mind that there is no *pure inductor*. There is always some resistance associated with any practical coil, because the conductors (wires) that make up the turns of the windings have resistance. It is this resistance that consumes power.

USES FOR INDUCTIVE REACTANCE

Inductive reactance can be very useful. It holds back the current in an ac motor and makes sure it does not draw too much from the line. A transformer is able to sit on the line without having a load on it due to inductive reactance. Filters also use the properties of coils to aid in their selecting the proper frequencies for cross-over networks in stereo speaker systems and for making sure the right station is tuned in on the television and the radio. In refrigeration, air-conditioning, and heating work, you will be primarily concerned with what inductance can do when it is present in a solenoid, relay, motor, or transformer.

TRANSFORMERS

So far you have looked at the terms inductance, mutual inductance, and self-inductance. Inductance is the opposition to current flow presented by a coil or inductor. A changing magnetic field produced by a changing electrical current can also be used in making what is referred to as a transformer. When the magnetic field produced by one coil is allowed to cut the windings of another coil nearby, it produces the transformer effect, or it introduces a voltage in the second nearby winding. This voltage and the amount of current available can be controlled by utilizing knowledge gleaned from years of experience and from some simple mathematical formulas.

Mutual Inductance in a Transformer

Two coils placed close to one another will allow the magnetic flux from each coil to be linked to the other. This is the basis of mutual inductance. A change in the current in one coil produces a magnetic field that cuts the turns of the second coil and induces a voltage. A mutual inductance thus exists between the two coils. This mutual linkage causes the effective inductance of each coil to be different from that of the self-inductance. The unit of mutual inductance is also the henry and is designated by the symbol L_M. A mutual inductance of 1 henry exists when a current change of 1 ampere per second in the first coil (primary) induces 1 volt into the second coil (secondary).

The mutual inductance between two coils depends on the closeness of the two coils, the way in which the coils are wound, and the reluctance of the mutual magnetic path. Three ways to increase the mutual inductance of two coils are shown in Fig. 8-10. Two coils that are coupled by a magnetic core have a higher mutual inductance. When the magnetic core makes a complete loop, as seen in Fig. 8-8,

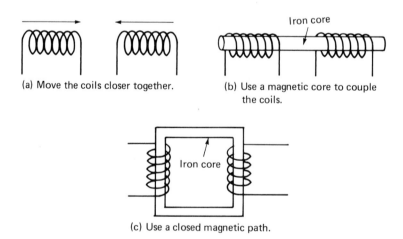

FIGURE 8-10 Three ways to increase mutual inductance.

nearly all the magnetic flux of one coil is coupled to the other coil, and the mutual inductance is maximum. The amount of flux linkage between two coils is called the coefficient of coupling.

Iron-Core Transformer

A device that transfers electrical power from one coil to another is called a *transformer*. Transformers may also change the value of voltage during the transfer. A transformer functions with no physical connection between the source and the receiving conductors. The principle used is mutual inductance.

Current flows into a transformer primary coil. This current creates a magnetic flux. The magnetic flux, in effect, couples the primary coil with the secondary coil. Voltage is induced in the secondary coil. The induced voltage may be varied by increasing or decreasing the magnetic field. The result of a transformer's operation is the induction of emf in the secondary coil. This is transferred to a conductor connected to the secondary coil.

Iron-core transformers use the mutual inductance principle to transform power between primary and secondary windings. An iron core provides a low reluctance path for the magnetic flux, and the coefficient of coupling between primary and secondary is so near unity (1) that it is considered as such in the designing of the transformer. The transformer is a highly efficient device since it has no moving parts; only current variations cause the transformer action to occur.

Changing magnetic fields are generated by ac so a transformer can be used to transfer ac power from primary to secondary (see Fig. 8-11). The two windings of a transformer are not connected, and the only means of transferring energy from the primary winding to the secondary winding is by way of the mutual coupling of a magnetic field.

Construction

Every transformer is built with at least two coils. One is a primary coil and the other is a secondary coil. The primary coil is called the input. This means that it brings electrical power into the transformer. The secondary coil is the output. It carries electrical power out of the transformer.

The transformer's iron core aids in the concentration of the magnetic field.

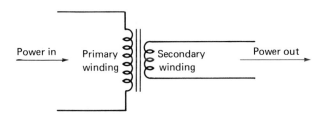

FIGURE 8-11 Step-down transformer.

The size of the core is the major determiner of the amount of current that can be drawn from the transformer, along with the size of the wire to handle the current. Wire size and core size determine the weight and physical size of the transformer. The higher the wattage rating or power output of the transformer, the larger the transformer is.

Refrigeration, heating, and air-conditioning circuits utilize a 24-volt transformer in most of the control circuits. This transformer steps down 120 volts to 24 volts. The National Electrical Code classifies this low-voltage control circuit as class II.

Types of Transformers

There are three basic types of iron-core transformers:

1. *Open core:* This type is used in power transformers (see Fig. 8-12).
2. *Closed core:* This type is more efficient than the open core type. Figure 8-13 shows that the flux path is contained within the core. This increases the strength of the magnetic field. The transfer of energy is improved.
3. *Shell core:* This is the most efficient type of transformer. Figure 8-14 shows the design characteristics. Note how the magnetic flux pattern containment improves the efficiency.

VOLTAGE TRANSFER

The transformer is a power transfer device. Power into the primary is transferred to the secondary. Also, power required by the secondary is reflected back as a power

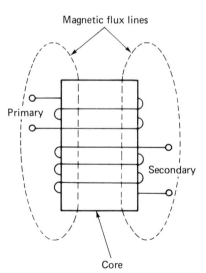

FIGURE 8-12 Open-core transformer.

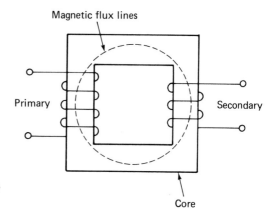

FIGURE 8-13 Closed-core transformer.

requirement to the primary. The rule is simple: *Power in is equal to power out,* less any losses.

Thus, if no load is connected to the secondary coil, the current flowing into the primary is virtually zero. The only current required is that amount necessary to support part of the losses. This is almost zero. Current flows in the primary only when the secondary is connected and there is output current to a load.

Step up, Step down

Transformers are either step up or step down. If input voltage is higher than output voltage, the device is known as a step-down transformer. In the step-up transformer, the output voltage is higher than the input.

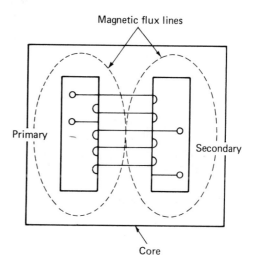

FIGURE 8-14 Shell-core transformer.

Turns Ratio. The input and output voltage relationship of a transformer depends on the turns ratio. The turns ratio describes the relationship of the windings of the primary and secondary coils. The first number in a given ratio is for the secondary. The second number is for the primary.

To find the turns ratio of a transformer, divide the number of turns of the secondary by the number of turns of the primary. For example, if there are 1000 turns in the secondary and 100 turns in the primary, divide 1000 by 100. This produces a ratio of 10 to 1. This is written as 10 : 1. The formula used to find the turns ratio is

$$\text{turns ratio} = \frac{1000}{100} = \frac{10}{1} \quad \text{or} \quad 10:1$$

If the transformer has a turns ratio of 10 : 1, the voltage ratio is 10 : 1. If 100 volts is applied to a 10 : 1 transformer primary, the secondary would have an output of 1000 volts.

There are other types of transformers that you should be familiar with. This will make it possible for you to make some substitutions, if needed, and give you some idea as to which transformers cannot be used for control circuits such as you will experience in refrigeration, air-conditioning, and heating equipment.

POWER TRANSFORMERS

Power transformers use multiple secondary coils. These can deliver a number of secondary voltages. Examples of the power transformer are shown in Figs. 8-15 and 8-16. These transformers are used for a wide variety of jobs. They may provide power for factories. They may power broadcast stations. Or they may be built into special equipment such as the power supply of a transmitter, television set, or electronics equipment. As you can see from the schematic, there are both step-up and step-down transformers. They have one primary coil with various ratios of secondary coils wound on top of the primary. They can be easily identified since they will have more than the usual four leads coming out of the case.

AUDIO FREQUENCY AND RADIO FREQUENCY TRANSFORMERS

Audio transformers change voltages for use in the audio range of frequencies. Audio frequencies are from 16 to 16,000 hertz. An audio transformer is shown in Fig. 8-17.

To deliver voltage at higher frequencies, no core is used. This is because an iron core consumes too much power above 16,000 hertz. Air-core transformers, therefore, are usually used in radio frequency circuits and are called *rf* transformers.

Electronics equipment has special miniature transformers to do certain jobs.

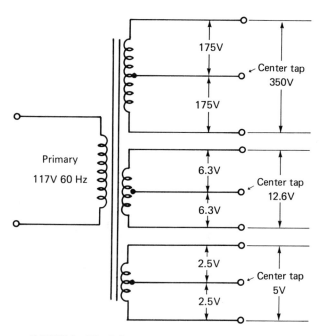

FIGURE 8-15 Schematic of a power transformer.

These are not usable on line frequencies such as 60 hertz for control circuits in heating and air-conditioning equipment.

AUTOTRANSFORMERS

Autotransformers have only one coil. Input and output come from a single coil. The amount of voltage output is determined by the position of the secondary tap (connection point). Placement of the secondary tap changes the turns ratio.

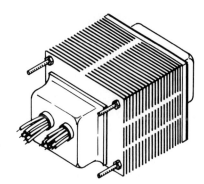

FIGURE 8-16 Power transformer used in electronic devices.

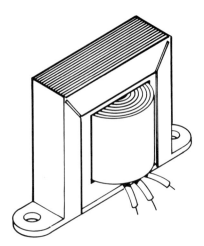

FIGURE 8-17 Audio-type transformer.

When an autotransformer is used to step up the voltage, part of the single winding is used as the primary. The entire winding acts as the secondary (see Fig. 8-18). Power is transferred from the primary to the secondary by the changing magnetic field that is concentrated by the core material. The turns ratio determines the voltage output.

An autotransformer does not provide isolation, as do the regular primary coil and secondary coil types that are not physically connected. The autotransformer will not serve to remove the ground found in most home and industrial wiring systems because the primary and secondary use the same turns. This type of unit can be made into a variable transformer. A variable transformer can produce differing output voltages. A slider arm is placed over the windings. The slider makes contact with points where the insulation on the wire has been removed. This type of variable transformer is called a *variac*.

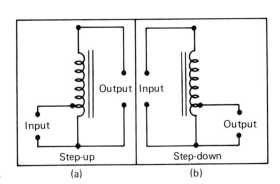

FIGURE 8-18 Autotransformer.

TRANSFORMER LOSSES

There are transformer losses, even though most of them can be minimized. Nevertheless, there are some that must be taken into consideration when designing a transformer for maximum efficiency. Three types of losses take power away from the output in a transformer and must be dealt with when the transformer is designed.

1. *Copper losses:* These are due to the resistance of the wire in the primary and secondary coils. Large size wire helps to minimize these losses.
2. *Hysteresis losses:* These are due to the properties of the iron core. Iron is slow to change polarity. The delay is known as hysteresis, or slowness to change properties (see Fig. 8-19). The magnetic domains (or molecules) of the iron core are facing in one direction. The domains (or molecules) are mixed in the middle illustration. This is in response to a change in magnetic fields. In the last illustration the domains (or molecules) are lined up in the opposite direction. This happens when the magnetizing force of ac is at its peak. Hysteresis losses in transformers are minimized by using silicon steel for the iron core. This type of metal offers very little opposition to changing polarity.
3. *Eddy currents:* These are very small but create losses in the transformer core. They are small currents created in the core when the magnetic fields change (see Fig. 8-20). Eddy currents generate heat. They flow in a direction opposite

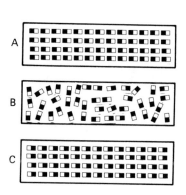

FIGURE 8-19 Hysteresis losses due to alignment of domains (molecules) in an iron core.

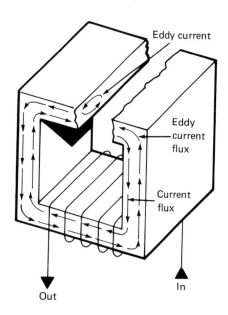

FIGURE 8-20 Eddy currents in a tranformer core.

to the current that induced them. The effect is to resist the flow of current in the core. Eddy currents can be minimized by using laminations. Laminating is the building of an object through the use of several layers of material. When this method is used, each lamination is varnished. Varnishing insulates the layers from each other. This increases resistance to eddy currents. Figure 8–21 shows how a solid core has high eddy current losses and a laminated core practically eliminates current losses.

INDUCTIVE CIRCUITS

Inductors have a definite effect on the operation of the circuits in which they are located. Inductors, remember, can be a coil of wire, the windings of an electric motor, or the ballast from a fluorescent lamp. Relay coils and solenoid coils have inductance, as do transformers. These devices have definite functions in circuits used in heating, air conditioning, and refrigeration. Their reaction to one another and to resistors in the same circuit is very important. They can cause some interesting problems that need to be understood for you to be able to properly and *safely* troubleshoot them.

The response of an inductor to dc is quite different from its response to ac. In ac circuits the inductor responds to the changing current by producing a self-induced voltage that tends to prevent any change in the amplitude of the current. In a dc circuit there is no inductive opposition to the current because the current is unchanging. Therefore, in a dc circuit the amount of current depends entirely on the total resistance in the circuit. This includes resistors, wiring, and coil winding resistance.

The dc current in an inductor is determined by the dc resistance. Assume that an ac and a dc voltage of the same value are applied to identical inductors. Will there be a difference in the values of the dc and ac currents? The current is higher

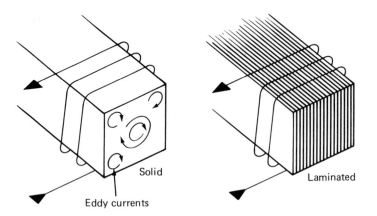

FIGURE 8–21 Solid core versus laminated core to eliminate eddy currents.

for the dc circuit because it is determined by the resistance of the coil windings. This value is usually very small. For the ac circuit, the current is determined by the impedance (impedance is the combination of resistance and reactance). Furthermore, as the frequency of the alternating current is increased, the coil current decreases because of the increasing reactance of the inductor.

Figure 8-22 shows how the coil reacts with dc applied. When the switch in the circuit is closed, a certain time is required for the magnetic field to expand to the maximum value determined by the current in the circuit. During this interval, the magnetic flux is changing, and the counter emf prevents the current from rising immediately to the final value determined by the resistance. The graph in Fig. 8-22B shows what happens. At time zero the switch is closed and the current begins to increase. When the switch is first closed, the rate of flux change is maximum, and, as shown by the voltage curve E_L, a maximum voltage appears across the coil. However as the current I rises, the rate of flux change slows down. The voltage across the coil decreases as the current rises.

This process continues until the current builds up to the maximum value determined by the dc resistance of the circuit. At this point, the magnetic field that has built up around the coil becomes stationary and unchanging, and the voltage drop across the coil, assuming the coil has no resistance, falls to zero. This is the steady-state condition. As long as the switch remains closed, this amount of current will be present in the circuit. The initial buildup period is brief, and in most dc circuits the time required to reach a steady state is very short. The delay can be witnessed in some circuits where it takes some time for the current to build up to its maximum value.

Keep in mind that the maximum amount of current is dumped back into the

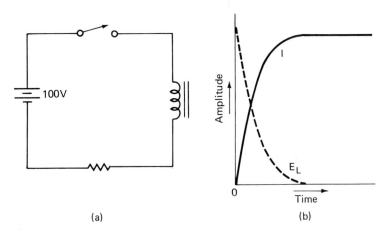

FIGURE 8-22 Current rises from zero to maximum. Maximum is determined by resistance in the circuit. The voltage across the coil drops from the applied voltage to zero.

circuit when the switch is opened. This collapsing magnetic field induces an emf of large amounts across the switch when it is opened. This is one reason why switches must be rated for inductive loads. This *inductive kickback* can cause an arc over and damage the switch if it is not properly rated to handle motors or inductive loads.

In ac resistor–inductor circuits, the current is determined by the impedance of the circuit. The higher the reactance and resistance of the circuit, the lower is the series current. The voltage drop across the inductor is determined by its reactance. Reactance is determined by both frequency and inductance. In a dc resistor–inductor circuit, a short but definite time is needed for the current to rise to its maximum value. This maximum current value is determined by the resistance of the circuit. The higher the resistance in the circuit, the lower the current is. However, assuming that the same two components are used in an ac and a dc circuit, the current present with the application of a dc voltage will be higher than the current present when an ac voltage of the same value is applied to the circuit. The dc voltage drop across the inductor in the dc circuit depends on the winding resistance of the inductor. If the winding resistance is assumed to be zero, the voltage drop across the inductor will be zero in a dc circuit.

Special Handling

In handling circuits of this nature, remember the inductor in the circuit can cause a mild shock when you use an ohmmeter to check its continuity. The ohmmeter has a dc power supply that can excite the windings of a motor, solenoid coil, or relay coil. If you get across the terminals being tested, it is possible to obtain a mild shock from a circuit you may assume to be de-energized.

UTILIZING THE INDUCTIVE DELAY

When a start capacitor is used in series with a compressor motor to cause it to start, it is taken out of the circuit within a few seconds. This is so the capacitor does not explode from having ac applied to it. The delay in energizing the relay to remove the capacitor from the circuit is caused by the inductance of the relay coil and the inductance of the motor winding. The delay is caused by inductive reactance and the normal delay caused by a coil to a changing current. This is one practical use for the inductive reactance of a coil.

REVIEW QUESTIONS

1. What are two other names for a coil?
2. What does the term inductance mean? What is its symbol?
3. What four things determine the inductance of a coil?
4. What is self-inductance?

5. What is mutual inductance?
6. What is meant by counter emf?
7. What does out of phase mean?
8. What is inductive reactance? What is its symbol?
9. List some uses for inductive reactance.
10. What determines mutual inductance?
11. What does a transformer do? How is it made?
12. How many coils does a transformer have?
13. What is meant by the turns ratio in a transformer?
14. What is the difference between an audio transformer and a radio frequency transformer?
15. How is an autotransformer different from a power transformer?
16. What is a variac?
17. What are the three types of transformers?
18. What are the three losses associated with a transformer?
19. What is meant by inductive kickback?

PERFORMANCE OBJECTIVES

Know what *capacitance* is.
Know how a capacitor works.
Know the three factors that determine a capacitor's capacity.
Know the basic unit of measurement for capacitance.
Know six types of widely used capacitors.
Know how an electrolytic capacitor is made.
Know what can cause an electrolytic capacitor to explode.
Know the meaning of WVDC.
Know why electrolytics are polarized.
Know what *capacitive reactance* is.
Know why a capacitor can cause a lagging voltage.

CHAPTER 9

Capacitors and Capacitive Reactance

To cause electric motors to start when single-phase power is applied, it is necessary to make it two phase or more. The split-phase motor was developed to utilize the single-phase power source, but it also needed an additional component in its start circuit to get it moving from a standing condition.

With the development of high-quality and high-capacity electrolytic capacitors, a variation of the split-phase motor known as the capacitor-start motor, has been made. Almost all fractional-horsepower motors in use today on refrigerators, oil burners, washing machines, table saws, drill presses, and similar devices are *capacitor start*. A capacitor motor has a high starting current and the ability to develop about four times its rated horsepower if it is suddenly overloaded. In this adaption of the split-phase motor, the start winding and the run winding have the same size and resistance value. The phase shift between the currents of the two windings is obtained by means of capacitors connected in series with the start winding. Capacitor-start motors have a starting torque comparable to their torque at rated speed and can be used in places where the initial load is heavy. One such application is a compressor in an air conditioner or refrigerator. A centrifugal switch is required for disconnecting the start winding when the rotor speed is up to about 25% of the rated speed. See Figure 9-1 for a disassembled capacitor-start motor. Note in Fig. 9-2, which also shows a capacitor-start motor, the centrifugal switch arrangement with the governor mechanism. The capacitor on this motor is shown on top in the motor frame. It is enclosed in the metal housing on top of the motor. More about motors will be presented in Chapter 11.

Now that you can see how important the capacitor is in air-conditioning, heating, and refrigeration circuits (especially motor circuits), you can see why you need

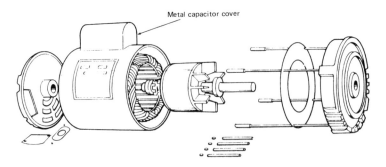

FIGURE 9-1 Exploded view of a capacitor-start motor.

a better understanding of how they operate in order to be able to make decisions about replacement, repair, and testing of the equipment using them.

THE CAPACITOR

A capacitor is a device that opposes any change in circuit voltage. The property of a capacitor that opposes voltage change is called *capacitance*.

Capacitors make it possible to store electric energy. Electrons are held within a capacitor. This, in effect, is stored electricity. It is also known as an electrostatic field. Electrostatic fields hold electrons. When the buildup of electrons becomes great enough, the electrical potential is discharged. This process takes place in nature when clouds build up electrostatic charges; their discharge is seen as lightning.

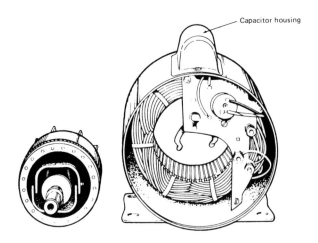

FIGURE 9-2 Single-phase stator and rotor.

HOW THE CAPACITOR WORKS 169

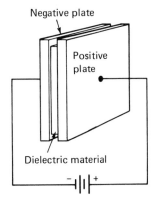

FIGURE 9-3 Simple capacitor.

Figure 9-3 shows the diagram of a simple capacitor. Two plates of a conductor material are isolated from one another. Between the plates is a dielectric. The dielectric material does not conduct electrons very easily. In fact, the dielectric is an insulator. The larger the surface area of the conductive material (the plates), the larger is the capacitor's capacitance.

HOW THE CAPACITOR WORKS

If a capacitor has no charge of electrons, it is uncharged. This happens when no voltage is applied to the plates. An uncharged capacitor is shown in Fig. 9-4. Note the symbol for the capacitor in this drawing. This is the preferred way to show a capacitor: a straight and a curved line facing each other. Figure 9-4B shows the switch closed to position 1. This causes current to be applied to the capacitor. A difference in potential is created by the voltage source (the battery). This causes electrons to be transferred from the positive plate to the negative plate of the capacitor. This transfer continues as long as the voltage source is connected to the two plates and until the accumulated charge becomes equal to the voltage of the battery. That is, charging takes place until the capacitor is charged.

In Fig. 9-4C, the voltage has been removed. The switch is open. At this point, the potential difference, or charge, across the capacitor remains. That is, there is still a surplus of electrons on the negative plate of the capacitor. This charge remains in place until a path is provided for discharging the excess electrons.

In Fig. 9-4D, the switch is moved to its third position. This opens the path for discharging the surplus electrons. Notice that the discharge path is in the opposite direction from the charge path. This demonstrates how a change in circuit voltages results in a change in the capacitor charge. Some electrons leave the excess or negative plate. They do this to try to keep the voltage in the circuit constant.

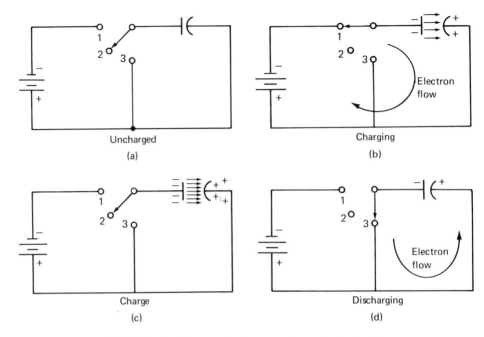

FIGURE 9-4 Charge–discharge action of a capacitor.

CAPACITY OF A CAPACITOR

The two plates of a capacitor can be made from almost any material. The only requirement is that the material must allow electrons to collect on its surface. The dielectric between the plates of a capacitor is an insulating material. Some dielectrics are air, vacuum, plastic, mica, wood, Bakelite, paper, and oil.

Three factors determine the capacitance of a capacitor:

1. Area of the plates
2. Distance between the plates
3. Material used as a dielectric

The area of the plates determines the ability of a capacitor to hold electrons. The larger the plate area, the greater the capacitance is.

The distance between the plates determines the effect that the electrons on the plates have on one another. There is an electrostatic charge around each electron. Electrons on the plates store energy when voltage is applied. Capacitance increases as the plates are brought closer together. Capacitance decreases as the plates are moved apart.

The material used as a dielectric determines the charge also. The thickness of the material means the plates are separated more or less by this dimension. But the type of material in the dielectric determines how much the electrostatic charges in the dielectric material react with one another. Some materials have the ability to increase the difference between the negative and positive plates.

BREAKDOWN VOLTAGE

The voltage at which the dielectric of a capacitor no longer functions as an insulator is the breakdown voltage. At this level, the capacitor permits the free flow of electrons through the dielectric. The result is known as a *shorted capacitor*. If the breakdown is partial, the result is a *leaky capacitor*. Thus, the dielectric strength of the material used is important to the functioning of a capacitor.

BASIC UNITS OF CAPACITANCE

The basic unit for capacitance is the *farad* (F). A 1-farad capacitor has stored 1 coulomb of electrons. It has a potential difference of 1 volt between its plates.

In most cases in electronics and electrical applications, the farad is too large a unit to be practical. It is broken down into the microfarad (μF) and the picofarad (pF), formerly called the micromicrofarad. The microfarad has been abbreviated UF, MF, MFD, Mfd. Today's standard abbreviation is μF.

The micromicrofarad was once used as the smaller unit of measurement of capacitance. Some older capacitors may be marked with a MMF, MMFD, UUF, or $\mu\mu$FD. Today's standard abbreviation is pF for picofarad.

The microfarad is one-millionth of a farad (0.000001). The picofarad is just what it says, one-millionth of one-millionth. The *pico* is today's accepted way of saying one-millionth of one-millionth. The Greek letter mu (μ) is used to represent micro. Micro is used universally as a prefix to indicate one-millionth.

TYPES OF CAPACITORS

Six general types of capacitors are the most widely used:

1. Air
2. Ceramic
3. Mica
4. Electrolytic
5. Paper
6. Tantalum

The electrolytic capacitor is marked with + and − and has polarity that must be observed when it is connected in a circuit. The other types do not need a polarity marking. Figure 9–5 shows an electrolytic capacitor that may be found in air-condi-

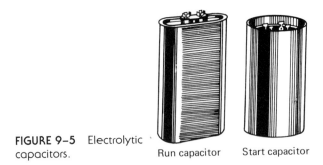

FIGURE 9-5 Electrolytic capacitors.

tioning, refrigeration, and heating applications. It is used as ac motor run and start capacitors (see Fig. 9-6).

Air Capacitors. Air capacitors have air for a dielectric. They are usually variable capacitors used in the tuning circuits of radios.

Mica Capacitors. Aluminum foil is used as the plate material in mica capacitors. Between the aluminum foil plates is a thin sheet of mica. Sometimes the mica is

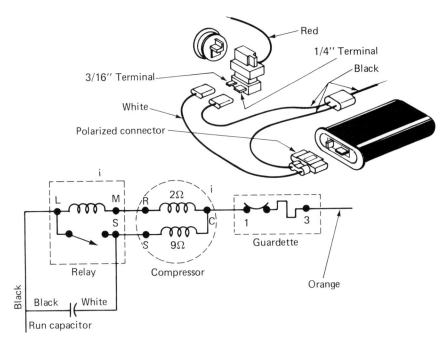

FIGURE 9-6 Electrical diagram of the capacitor-run type of refrigerator compressor. Note that the capacitor has a polarized plug so that it cannot be connected incorrectly.

FIGURE 9-7 Mica capacitors.   **FIGURE 9-8** Paper capacitors.

sprayed with a conducting paint. The paint then forms the plate on one side of the mica. Mica capacitors are usually sealed in Bakelite or some type of plastic (see Fig. 9-7).

Paper Capacitors. Aluminum is also used as the plate material in paper capacitors (see Fig. 9-8). However, the plates are separated by a paper dielectric. The materials (paper and aluminum) are rolled into a cylindrical shape. A wire is connected to alternate ends of the foil and it is encased in plastic.

Ceramic Capacitors. Ceramic dielectric materials make high-voltage capacitors. They have very little change in capacitance due to temperature changes. These small capacitors usually consist of a ceramic disc coated on both sides with silver. They are made in values from 1 picofarad up to 0.05 microfarad. Breakdown voltages of ceramic capacitors run as high as 10,000 volts and more.

Oil-filled Capacitors. Oil-filled capacitors are paper capacitors encased in oil. They are sometimes referred to as bathtub capacitors (see Fig. 9-9). The main advantages of these capacitors are sturdy construction and high voltage breakdown ratings. They are used in places where grease and oil are likely to be encountered.

ELECTROLYTIC CAPACITORS

Electrolytic capacitors are very important to people working in the air-conditioning, refrigeration, and heating field.

There are two types, wet and dry. The wet type uses a liquid electrolyte. It is

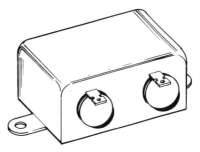

FIGURE 9-9 Oil-filled capacitor.

rather large in physical size. But it has high capacitance. Wet electrolytics are generally used in transmitters and other large, stationary electronic equipment.

The dry electrolytic capacitor is used for capacitances above 1 microfarad. It can be made in sizes up to 1 farad. These units can be kept rather small in physical size for such large capacitances. This is done by using an oxide film as a dielectric (see Fig. 9-10). Very large sizes are used in computer power supplies for filtering the dc power supply.

Making an Electrolytic Capacitor

The way an electrolytic capacitor is made is very important because it also affects the way in which the capacitor is used in a circuit. To produce a capacitor of this type, dc voltage is applied to the electrolytic capacitor as part of the manufacturing process. This produces an electrolytic action. As a result, a molecule-thin layer of aluminum oxide with a thin film of gas is deposited. This gas is at the junction between the positive plate and the electrolyte.

The oxide film is a dielectric. There is capacitance between the positive plate and the electrolyte through the film. The negative plate provides a connection to the electrolyte. This film allows many layers of film to be placed in a case. Thus, large capacitances can be placed in smaller containers since the insulating film between the plates is so thin that the plates are very close together. And, remember, the closer the plates are the greater the capacitance.

Connecting Electrolytics

Electrolytics have polarity (− and +). If these capacitors are not connected in a circuit properly, the oxide film is not formed. If connections are improper, no capacitance is available. Reverse electrolysis forms a gas when the capacitors are connected to opposite polarity. These capacitors become hot and may explode.

Electrolytic capacitors may dry up and become *leaky*. These cause all kinds of

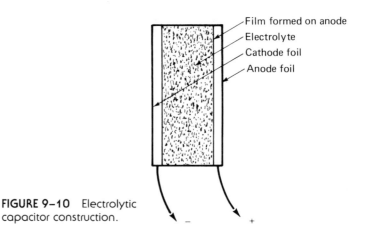

FIGURE 9-10 Electrolytic capacitor construction.

problems since they may re-form to their rated capacitance once the power has been turned on for some time. So the problems they present may be intermittent. These are the hardest types to pin down when troubleshooting. Electrolytic capacitors should be used in circuits in which approximately 75% of their working voltage is available. This keeps the capacitor formed. If a refrigerator or air-conditioning unit has been left out of service for a year or more, it is possible for the electrolytic capacitor to present problems. It may have dried up or become leaky.

Nonpolarized electrolytic capacitors are available. They do not have − and + polarity. They use two capacitors in series in the same case and connected in what is called a back-to-back configuration. This means that the two negative terminals are connected internally and the positive terminals are brought out of the case to be connected in the circuit. Nonpolarized electrolytics are used for ac circuits since the normal electrolytic capacitor would explode when ac is connected (see Fig. 9-11).

WORKING VOLTAGE, DIRECT CURRENT (WVDC)

The maximum safe operating (working) voltage of a capacitor in a dc circuit is identified as working volts, direct current, or WVDC. When higher voltages are applied, the dielectric will deform and allow electrons to pass through, thus ruining the capacitor.

CAPACITIVE REACTANCE

The degree to which a capacitor opposes the flow of current is called capacitive reactance (X_C). The capacitive reactance of a circuit is determined by capacitance and frequency. These two factors are used in a formula to determine the capacitive reactance of any given circuit. The formula is

$$X_c = \frac{1}{2\pi fc}$$

In this formula, X_C is the symbol for capacitive reactance in ohms, C is the symbol for capacitance in farads, and f is the frequency in hertz.

You can see, then, that the changing of the frequency from 60 to 50 hertz would make a difference in the operation of the capacitor. To make the capacitors usable for both frequencies, they have been designed for about 55 hertz so they will operate on both frequencies. This means that the electrolytic capacitors you use for

FIGURE 9-11 Back-to-back connection of electrolytics to provide a nonpolarized capacitor.

motors have to be properly sized for 50/60 hertz to prevent overheating. In most instances today, the capacitors are made for 50/60 hertz operation.

Note also that the size of the capacitor is very important with respect to the size of motor. In most instance the electrolytics capacitors will be marked with a range of capacitances:

88–106 μF at 110/125 volts, 50/60 hertz

108–130 μF at 110/125 volts, 50/60 hertz

540–648 μF at 110/125 volts, 50/60 hertz

88–106 μF at 220/250 volts, 50/60 hertz

108–130 μF at 220/250 volts, 50/60 hertz

This type of electrolytic is equipped with two quick-disconnect terminals or some variation (see Fig. 9-12).

The main use for electrolytic capacitors in motor or compressor circuits is for providing starting torque for the device. If a motor must start under load, such as a compressor, it is usually a capacitor-start type.

CAPACITOR CAUSES A LAGGING VOLTAGE

In a circuit with a capacitor only, the voltage lags, or follows the current by 90°. This is because a purely capacitive circuit has no resistance (a condition that is never found in a practical circuit). In a purely capacitive circuit, current and voltage are said to have a phase relationship. The phase difference is 90° (see Fig. 9-13).

The current and voltage relationships of capacitance and inductance are found in circuits with motors providing the inductance. By the proper use of capacitance, it is possible to bring the circuit back *into phase* and reduce the current drain from the power source. Run capacitors have this effect since they are left in the circuit at all times. This aids in keeping down the operating costs for electricity.

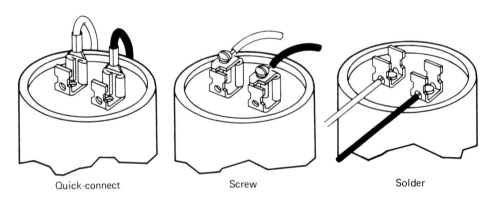

Quick-connect   Screw   Solder

FIGURE 9-12 Three ways to connect electrolytics in a circuit.

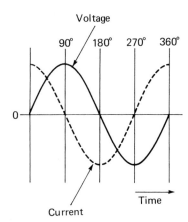

FIGURE 9-13 Voltage and current waveforms in a purely capacitive circuit.

CHECKING CAPACITORS

Start capacitors and run capacitors have a tendency to become open, leaky, and short. These conditions can cause a variety of symptoms that the troubleshooter must become familiar with to quickly recognize what to do to correct a problem. To check for these conditions, it is necessary to use an ohmmeter.

1. Set the ohmmeter to read its highest resistance range.
2. Short the leads together and adjust the zero-adjust knob until you get a zero reading on the meter. (The battery in the meter will charge the capacitor and, in so doing, a keen observer can detect any malfunctioning of the capacitor).
3. Use a piece of insulated wire with insulation removed at the ends to short out the capacitor by placing it across the capacitor terminals. You may get a spark and you may not. Do this five times to make sure you have discharged the capacitor. If the capacitor has a resistor across its terminals (see Fig. 9-14), remove one end by desoldering or pulling it loose from the connector. This must be done to prevent damage to the meter movement.
4. Place the meter probes across the capacitor and hold them there for at least

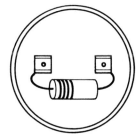

FIGURE 9-14 Bleeder resistor across capacitor terminals.

15 seconds. Watch the action of the meter at this time. The needle on the meter should swing all the way to the right or zero. Then it will gradually move back toward infinity and stop. This indicates that the capacitor is taking a charge and is probably all right to put in the circuit.

If the needle goes to zero and stays, the capacitor is shorted. Replace it. There are other ways of finding if a capacitor is doing its job. You may also

TABLE 9-1 Capacitor Troubles

Trouble	Probable Cause	Remedy
Compressor will not start. However, it hums intermittently. Cycles with the protector.	Start capacitor open.	Replace start capacitor.
Compressor starts. Motor will not speed up enough to have start winding drop out of circuit.	Start capacitor weak. Run capacitor shorted.	Replace the capacitor. Disconnect run capacitor and check for short.
Compressor starts and runs. However, it cycles on the protector.	Run capacitor defective.	Check capacitance. Replace if found defective.
Start capacitors burn out.	Short cycling.	Reduce number of starts. They should not exceed 20 per hour.
	Prolonged operation with start winding in circuit.	Reduce the starting load. Install a new crankcase pressure limit valve. Increase low voltage if this is found to be the condition. Replace the relay if it is found to be defective.
	Wrong relay or wrong relay setting.	Replace the relay.
	Wrong capacitor.	Check specifications for correct sized capacitor. Be sure the MFD and WVDC are correct for this compressor
	Water shorts out terminals of the capacitor.	Place capacitor so terminals will not get wet.
Run capacitors burn out. They spew their contents over the surface of anything nearby. This problem can usually be identified with a visual check.	Excessive line voltage.	Reduce line voltage. It should not exceed 10% of the motor rating.
	Light load with a high line voltage.	Reduce voltage if not within 10% overage limit.
	Voltage rating of capacitor too low.	
	Capacitor terminals shorted by water.	Replace with capacitor of correct WVDC. Place capacitor so the terminals will not get wet.

check from each terminal of the capacitor to its metallic case. If you get a zero reading, replace the capacitor. The meter should read infinity from each terminal to the case to be usable.

A troubleshooting guide for capacitors is given in Table 9-1.

REVIEW QUESTIONS

1. What is capacitance? What is its symbol?
2. How does a capacitor work?
3. What three factors determine the capacity of a capacitor?
4. What is the basic unit of measurement for capacitance?
5. What is a picofarad? What is a microfarad?
6. What are six types of widely used capacitors?
7. How is an electrolytic capacitor made?
8. What causes electrolytic capacitors to explode?
9. What does working voltage mean?
10. What is polarization and how does it affect electrolytic capacitors?
11. What is capacitive reactance? What is its symbol?
12. How does a capacitor cause a lagging voltage?

PERFORMANCE OBJECTIVES

Know the meaning of *impedance*.
Know what a vector is and how it works.
Know what the term *kilowatt-hour* means.
Know what is meant by *true power*.
Know the difference between *apparent power* and *true power*.
Know what *reactive power* is.
Know how most electrical power is generated.

CHAPTER 10

Single-Phase and Three-Phase Alternating Current

Alternating current affects capacitors and inductors differently than direct current. The reactance of inductors and capacitors to ac creates a phase shift between the voltage and current in the circuit. This phase shift causes the kilowatt-hour meter that measures the power consumed in a home, business, or industrial plant to read higher than the power that is actually being used. Inasmuch as compressors for air conditioners and refrigerators are motors, they are also known as inductors and cause a phase shift. In some cases it is possible by using the right combination of capacitance to lower the power being measured on the kilowatt-hour meter and, of course, lower the cost of operation of the unit. This is why service technicians need to know about power factor and how it affects the equipment they work on.

RESISTANCE, CAPACITANCE, AND INDUCTANCE

The total opposition to the flow of current within a circuit is called *impedance* (Z). It impedes or resists the flow of current. Since it impedes the flow of current in a circuit, impedance is measured in ohms.

In dc circuits, the only opposition is resistance. In ac circuits, opposition to voltage and current comes from a combination of factors. These include resistance (R), capacitance (C), and inductance (L).

The effect of combining these elements (R, L, and C) in a circuit is best examined in steps. Start with resistance. Then add inductance and/or capacitance. When combined in a circuit, a resistor and inductor and/or capacitor produce impedance. Impedance is a combination of effects. It consists of resistance and reactance provided by an inductor and/or a capacitor.

182 SINGLE-PHASE AND THREE-PHASE ALTERNATING CURRENT

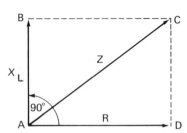

FIGURE 10-1 Vectors showing impedance.

Keep in mind that voltage leads current across an inductor and lags current across a capacitor. This is a *phase angle* of 90°. The effect on current is of special interest in this situation. The reactance effect is 90° out of phase with the resistive effect. The voltage and current, remember, are in phase in a resistor. To combine these, it is necessary to obtain the *vector sum* of the two quantities (resistance and capacitive reactance).

Impedance is the vector sum of resistance and reactance. A *vector is a line segment* used to represent a quantity that has both direction and magnitude. Vectors are used to represent current. A single vector line, or curve, can show two dimensions of current. It can show the direction of flow and it can also show the magnitude or amount of current flowing. A *vector sum* is a line representing the total of two or more vectors. Impedance is stated in terms of a vector sum (see Fig. 10-1). In this illustration you can see how vectors represent X_L and R. The Z is used to represent the *sum* of the two acting together to produce the impedance. By measuring the angle made by *CAD,* you can also find the phase angle produced by the inductor and resistor being in the same circuit. The phase angle is directly related to the power factor. Power factor is important in ac circuits because it tells the ratio of power being consumed to power that appears to be consumed and is actually read by wattmeters. More about power factor will be presented in this chapter as we develop the concepts needed to understand it.

POWER

Power is represented by the letter *P.* It is measured in watts. We also see it represented as kilowatts (kW). The kilo means 1000. Therefore, kilowatts means 1000 watts. When power is consumed at home or in industry for a period of a month, it is likely to be measured in terms of kilowatts, rather than in watts.

An ac circuit consumes power differently than a dc circuit or an ac circuit with resistance only. The addition of a capacitor or inductor makes a difference in the actual power consumed. When power is consumed in a dc circuit or an ac circuit with resistance only, the power is found by $P = E \times I$. Or the power is equal to the product of volts and amperes. Once a capacitor or inductor is added to an ac circuit, all this changes because the difference between voltage and current or phase shift has to be taken into consideration to get an accurate reading of what is actually consumed in the way of power. Power that is actually consumed is referred to as *true power.* True power is that which a dc circuit or ac *resistive* circuit consumes.

In a purely resistive circuit, energy is dissipated in the resistance as heat, and the resistance does not care in which direction the current is moving at any given instant. As a result, the power cycle looks like the upper waveform in Fig. 10-2. This resistive power cycle is always positive, indicating that power is always being drawn from the circuit. The average power for the cycle is one-half of the maximum value.

In a purely inductive or capacitive reactance, the situation is quite different. During parts of the voltage cycle, when the magnetic field is building up around the inductor in either direction or when the capacitor is being charged in either direction, power is being drawn from the circuit. This energy is stored in the magnetic field of the inductor or the electrostatic field of the capacitor. During the remaining

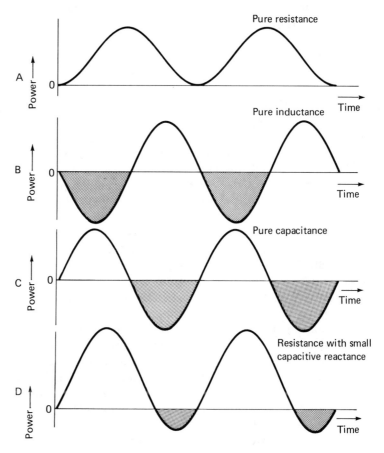

FIGURE 10-2 Power waveforms.

parts of the cycle, the magnetic field of the inductor collapses or the capacitor discharges, returning the stored power to the circuit. In these cases, we have a reactive power cycle that is sinusoidal about the zero line, with twice the voltage cycle frequency. The average power consumed for the cycle is zero. The capacitive cycle is 180° out of phase with the inductive cycle, as seen in Fig. 10-2B and C.

The power cycle in any real circuit lies somewhere between the extremes of pure resistive and pure reactive, with more power being drawn from the circuit than is returned to it (see Fig. 10-2D). The circuit is predominantly resistive but slightly capacitive. A slightly inductive power cycle would be similar, except that the negative parts of the cycle would precede the positive parts.

POWER FACTOR

The measure of the relative reactivity of an ac circuit is its power factor. Power factor is defined as the net resistance (R) divided by the total impedance (Z) of the circuit, or

$$PF = \frac{R}{Z}$$

Power factor is given in both percentage form and in decimal form. That is, 100% is 1.00, or 90% is 0.900.

In a purely resistive circuit, the impedance is made up only of resistance. That means the power factor is unity (100%). In a purely reactive circuit, the impedance comprises only reactance with no resistance, and the power factor is zero (0%). To find the apparent power (AP), you multiply the voltage E by the current I. This produces what is called volt-amperes (VA). Note this is different from watts. The difference in volt-amperes and watts is significant. The term watt is used to designate a purely resistive circuit or one with only dc resistance. The term volt-ampere designates that the circuit is ac and that there is an inductor or capacitor in the circuit. In other words, it is a reactive circuit when it has volt-amperes. Volt-ampere is also used when referring to the capacity or ability of a transformer to supply a certain amount of volts and amperes. For instance, a door chime transformer may put out 16 volts but is limited to a maximum of 1 ampere. This means that the volt-ampere rating of the transformer is 16 VA: $16 \times 1 = 16$. If you want to have two door bells or chimes operate on a transformer, it must be larger in terms of current available. Thus, if you buy a transformer with the ability to furnish 2 amperes and 16 volts, you have to ask for one that is rated at 32 volt-amperes: $16 \times 2 = 32$.

If you want to know the true power (W) instead of the volt-amperes (VA), you have to multiply by the power factor (pf). In other words, if you have 150 VA and a power factor of 0.5, you will have a true power of 75 watts: $150 \times 0.5 = 75$.

Reactive power averages out to zero. It can be a nuisance in most cases for it puts an instantaneous load on the equipment for part of the cycle, just as if it were a resistive power demand. For this reason, large industrial consumers of power from ac sources will deliberately introduce the opposite kind of reactance into a circuit. That is, they will put a capacitor into an inductive load to reduce the level of reactive power and bring the power factor closer to 100%. This is also the reason why ac generators are always rated in terms of the total apparent power, in kilovolt-amperes (kVA), that they can produce. For loads, on the other hand, it is important to know the two individual components of the apparent power: reactive power in kilovolt-amperes reactive (kVAR) and the true power in kilowatts (kW).

By adding a capacitor in parallel with an electric motor, the power factor can be improved. This is why some large consumers of electrical power try to reduce the power factor as much as possible. It means they have to pay for the apparent power. Apparent power is always more than true power when there is an inductor or capacitor in the circuit.

Keep in mind, though, that the capacitors used in compressor circuits are not there for power factor correction, but for improving starting torque. The power factor is a secondary consideration. Note that some motors are capacitor start and capacitor run. There are some advantages to having a capacitor in the run circuit. These advantages will be discussed in Chapter 12.

Power factor considerations are not the responsibility of the service technician. Power factor is presented here to give you a better understanding of all aspects of the electrical circuits you will encounter while working on refrigeration and air-conditioning equipment.

DISTRIBUTING ELECTRIC POWER

Most electric power is generated as three phase (3ϕ). It is stepped up to 138,000 volts, 345,000 volts, or even to 750,000 volts. The frequency is 60 hertz. Sometimes, however, 25 hertz ac is generated for use by some consumers who may have older equipment that uses this frequency. Motors and other equipment using this power frequency would be very expensive to replace. However, most 25-hertz equipment, when worn out, is replaced with 60-hertz equipment. A separate distribution system is necessary to send 25-hertz power to its destination. This is expensive, but we mention it here so that you will be aware of the fact that it exists in some places where there is a nearby generating plant.

Don't make the mistake of connecting 60-hertz equipment to 25-hertz power (see Fig. 10–3).

POLYPHASE

The use of more than two wires for a circuit delivering electrical energy offers advantages in economy and in the ways in which the energy can be utilized. As a result, several types of circuits using more than two conductors have been developed and

186 SINGLE-PHASE AND THREE-PHASE ALTERNATING CURRENT

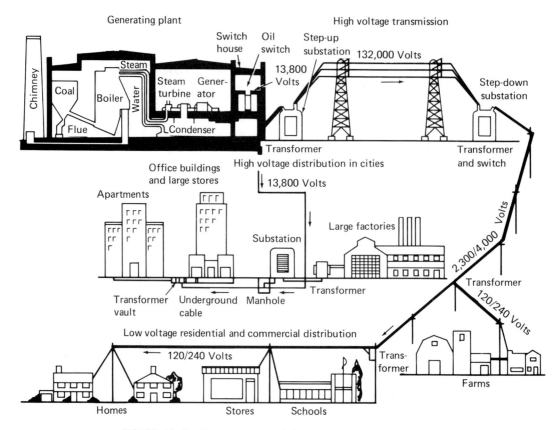

FIGURE 10-3 Generation and distribution of electrical power.

are in common use. These may be either single phase or polyphase (poly means more than one).

In a single-phase system having more than two wires, the voltage between any two pairs of wires will be substantially in phase with the voltage between any other pair of wires. Voltages added in series may be added numerically (taking into account polarity). In the single-phase, three-wire circuit shown in Fig. 10-4, for instance, if the line-to-neutral voltages are each 120 volts, the line-to-line voltage is 240 volts.

In a polyphase system, such voltages are usually not in phase. In a three-phase system (by far the most common polyphase system), three voltages are generated that are equal in magnitude, but 120° apart in phase. Plotted against time, these voltages look like Fig. 10-5. Note that all the voltages are equal in magnitude, but lag each other by 120°.

Figure 10-6 is a three-wire network being metered with a kilowatt-hour meter. Note how the voltages are distributed. In a conventional three-phase, three-wire

POLYPHASE **187**

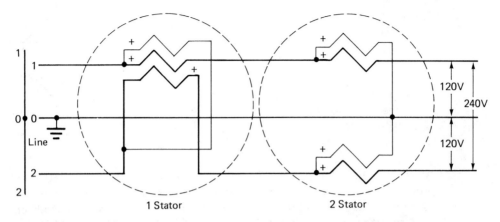

FIGURE 10-4 Single-phase, three-wire, 240-volt kilowatt-hour meter hookup. The zigzag lines represent the meter coil.

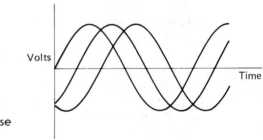

FIGURE 10-5 Three-phase relationships.

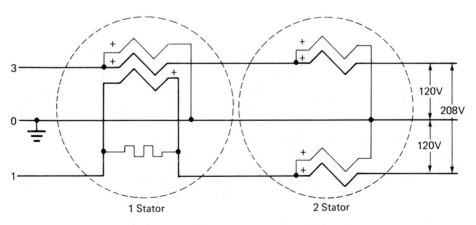

FIGURE 10-6 Three-wire network, 208 volts.

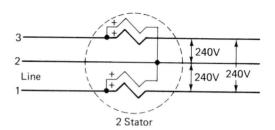

FIGURE 10-7 Three-phase, three-wire, meter hookup (delta).

system, all line voltages are equal in magnitude, usually at 240 or 480 volts; or for primary metering, at higher distribution voltages, a two-stator meter is required, which is accurate regardless of voltage unbalance. Such a system, usually employed for large commercial or industrial loads, may or may not be grounded. If one line is grounded, the current coils must be in series with the ungrounded lines (see Fig. 10-7).

If a center tap is brought out from one leg of a three-phase delta, the system becomes a three-phase, four-wire delta, such as that frequently used to supply polyphase power at 240 volts for motors and 120 volts, single-phase for lighting. The three-stator kilowatt-hour meter (Fig. 10-8) is rarely used, partly because of the complicated testing procedure. Instead, the two-stator, four-wire delta meter shown in Fig. 10-9 is almost universally used in this service.

Although a four-wire delta system is normally used only for 240/120-volt service, the three-phase (3ϕ), four-wire system covers a wide range of voltages from 208/120-volt service up to the highest transmission voltages (see Fig. 10-9). The neutral is usually grounded. It is measured with either a three- or two-stator meter, depending on the importance of the load and the degree of accuracy required. Note the distribution voltages between any two of the conductors.

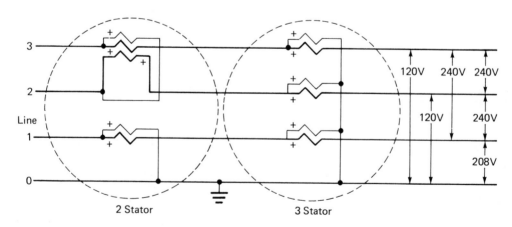

FIGURE 10-8 Three-stator meter (rarely used).

POLYPHASE **189**

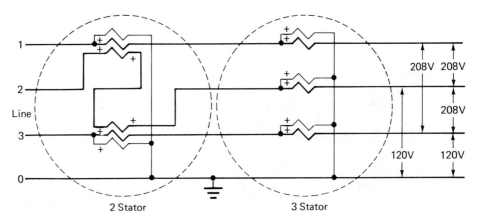

FIGURE 10-9 Four-wire wye system being metered.

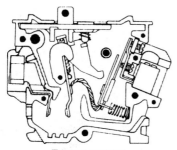

Tripped position
- Contacts open — no current flow
- Handle stationary when tripped

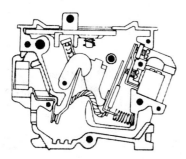

"On" position
- Contacts closed — current on
- Handle in "On" position (shows "On")

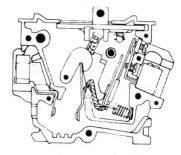

"Off" position
- Contacts open — no current flow
- Handle in "Off" position (shows "Off")

To restore service when fault is cleared you simply move operating handle to "Off" position and then to "On".

FIGURE 10-10 Cutaway view of a circuit breaker.

190 SINGLE-PHASE AND THREE-PHASE ALTERNATING CURRENT

These meter hookups are shown to aid in identifying the type of power supplied to various equipment.

CIRCUIT BREAKERS

A circuit breaker is a device that, after breaking a circuit, can then be reset by turning the handle to past the off position and then to on (see Fig. 10-10). The circuit breaker is mounted in a distribution box by snapping it into place. The hot wire (black or red) is attached by inserting it under the screw and tightening. A knockout blank in the distribution box must be removed to allow for the handle and top of the circuit breaker to be exposed (see Fig. 10-11).

A number of types of circuit breakers are used in home and industrial applications. However, most of them snap in place much as shown in Fig. 10-12. This illustration shows the snap in action without the sides or top of the box in place. Keep in mind that the circuit breakers are snapped into place in the panelboard. The wires are connected before the circuit breakers are snapped into place. Make sure the circuit breaker is in the off position while installing or reinstalling if it becomes necessary. Note there is a difference in the three types of circuit breakers shown here. Types A and C must have a notch so the formed rib fits properly. Type B can be used only where there is no formed rib in the mounting plate.

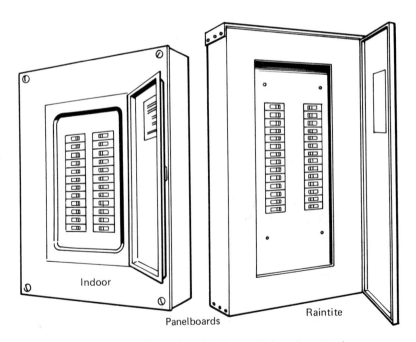

FIGURE 10-11 Circuit breaker box with breakers in place.

National Electrical Code: Section 384-15

"A Lighting and Appliance Branch Circuit Panelboard shall be provided with physical means to prevent the installation of more overcurrent devices than that number for which the Panelboard was designed, rated and approved."

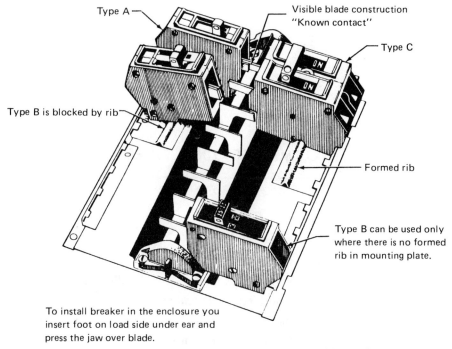

FIGURE 10-12 Circuit breakers are snapped into place in the panelboard.

REVIEW QUESTIONS

1. What is impedance?
2. What is meant by vector sum?
3. What is meant by the term kilowatt-hour?
4. What is meant by the term true power?
5. How do you measure the relative reactivity of an ac circuit?
6. What is the difference between apparent power and true power?
7. What is reactive power?
8. How is most electrical power generated?

PERFORMANCE OBJECTIVES

Know the meaning of the word *semiconductor*.
Know the two materials used for semiconductors.
Know how diodes are made and used.
Know how the SCR is used in air-conditioning and refrigeration circuits.
Know how an integrated circuit is produced and used.
Know how a thermister works.
Know how a humidity element works.
Know how a bridge circuit is used in control circuits.
Know how a sensor operates.
Know what an actuator does.
Know what a differential amplifier is used for.

CHAPTER 11

Solid-State Controls

The word *semiconductor* identifies a type of electronic device. Transistors and diodes are semiconductors. These units are part of the field of electronics. They have been used over the past few years in every aspect of control circuits on every type of equipment. Electronic controls for air-conditioning, refrigeration, and heating equipment have been upgraded to use transistors, diodes, and other types of semiconductor devices (see Fig. 11-1).

SEMICONDUCTOR PRINCIPLES

Semiconductor technology is usually called *solid state*. This means that the materials used are of one piece. This is in contrast with the vacuum tube, which consisted of a series of assembled parts. Of course, the substances from which semiconductors are made are not really solid. They have atomic structures consisting largely of empty space. The spaces are essential for the movement of electrons.

Many materials can be classified as semiconductors, but only two are used extensively for electronic circuits. They are *silicon* and *germanium*. Germanium was used originally for diodes and transistors, but has since been largely replaced by silicon.

Both silicon and germanium are hard, crystal-type materials that are very brittle. They can be made pure to 99.999999%. Doping agents or impurities are added in controlled amounts. Impurities used may be boron, aluminum, indium, gallium, arsenic, and antimony.

By doping silicon, it is possible to allow the electrons added by doping (in N-type material) to move easily. In P-type material, the holes or places where an

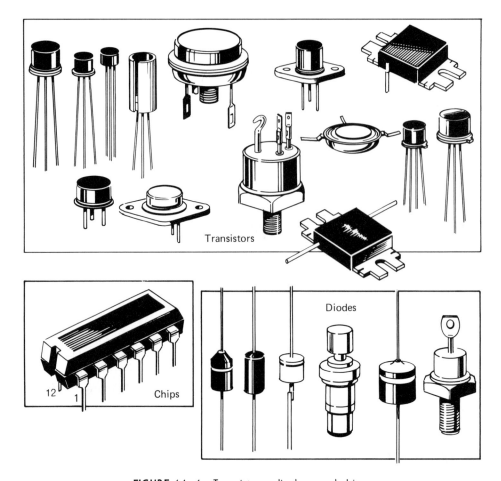

FIGURE 11-1 Transistors, diodes, and chips.

electron should be are moved along by the voltage applied. The addition of arsenic or antimony makes an N-type semiconductor material. The material will have an excess of electrons. Electrons have a negative charge.

When gallium or indium is used as the impurity, a P-type semiconductor material is produced. This means it has a positive charge, or is missing an electron.

DIODE

The semiconductor diode is used to allow current to flow in only one direction. It can be used to change alternating current to direct current. It is made by using N-type and P-type semiconductor materials.

When N-type and P-type materials are joined, they form a diode (see Fig. 11-2). The diode is also called a rectifier since it can change (rectify) ac and make it

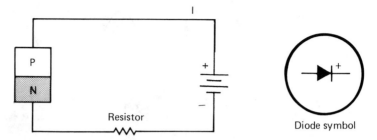

FIGURE 11-2 PN diode in a circuit. Note the symbol for a diode.

dc. The PN junction (diode) acts as a one-way valve to control the current flow. The forward or low-resistance direction through the junction allows current to flow through it. The high-resistance direction does not allow current to flow. Thus, only one-half of an alternating current hertz is allowed to flow in a circuit with a diode (see Fig. 11-2). The diode used in this circuit is forward biased and allows current to flow. Figure 11-3 indicates the arrangement in the reverse-bias configuration. No current is allowed to flow under these conditions. Also, note the polarity of the battery.

Diodes are used in isolating one circuit from another. A simple rectifier circuit is shown in Fig. 11-4. The output from the transformer is an ac voltage, as shown in Fig. 11-5. However, the rectifier action of the diode blocks current flow in one-half of the sine wave and produces a pulsating dc across the resistor (see Fig. 11-5B).

A number of factors affect the voltage and current ratings of a PN junction diode. As you work with them, you will learn that they have ratings for peak and inverse ac currents. This information is available for both germanium and silicon diodes in transistor handbooks produced by their manufacturers.

Figure 11-6 shows a number of specialized diodes. This illustration demonstrates that diodes are designed for specific jobs. Both internal and external con-

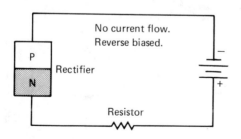

FIGURE 11-3 Reverse-biased diode in a circuit. Note how this differs from Figure 11-2.

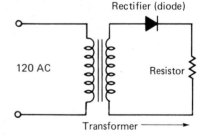

FIGURE 11-4 Circuit showing how a diode is used to rectify ac to produce dc.

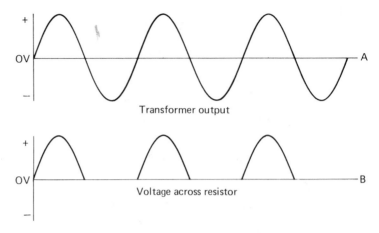

FIGURE 11-5 Results of the rectifier circuit. Transformer output (ac) is changed to dc.

struction are determined by circuit requirements. For example, Fig. 11-6A shows diodes used for radar and computer circuits. The diodes in Fig. 11-6B are designed to carry large currents. So their cases are heat conductors. As the diodes become warm, the generated heat is transferred to the air. Figure 11-6C shows zener diodes. These units protect sensitive meter movements and regulate voltage.

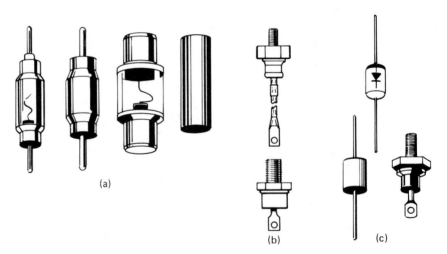

FIGURE 11-6 Diodes for special purposes. (A) Radar and computer circuits. (B) Large-current circuits. (C) Low-current circuits.

SILICON-CONTROLLED RECTIFIERS

The silicon-controlled rectifier (SCR) is a specialized type of semiconductor used for control of electrical circuits. This is a four-layer device. The structure can be either NPNP or PNPN.

An SCR conducts current in a forward direction only. The symbol for an SCR is shown in Fig. 11-7. Current always flows through an SCR from the cathode (C) to the anode (A). The illustration indicates that the SCR also has a gate (G).

The function of an SCR is shown in the circuit diagram in Fig. 11-8. The typical use of an SCR is for a controlled circuit. Examples include a light dimmer or a speed control for a motor. This type of circuit is shown in Fig. 11-8. The resistor in the circuit, R_1, is a rheostat, or adjustable resistor. This is used to control the amount of voltage delivered to the gate of the SCR. The more voltage delivered, the greater the flow. Thus, adjusting the rheostat will control the circuit. If the circuit illuminates a lamp, lowering the voltage to the rheostat dims the bulb. If the load is a motor, its speed is lowered. Figure 11-9 shows typical SCRs.

TRANSISTORS

Transistors are made from N- and P-type crystals. Once joined, the two different types of crystals produce junctions. Transistors are identified according to emitter junction and collector junction. Thus, they are either PNP or NPN types.

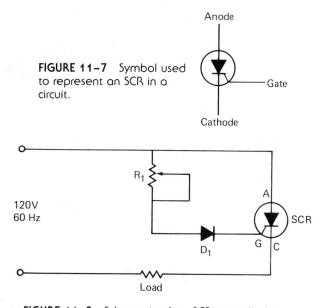

FIGURE 11-7 Symbol used to represent an SCR in a circuit.

FIGURE 11-8 Schematic of an SCR-controlled circuit.

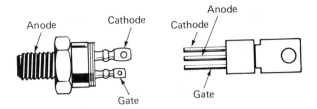

FIGURE 11-9 Two types of SCRs. Both types are used in control circuits depending on the amount of current involved.

A PNP transistor is formed by a thin N region between two P regions (see Fig. 11-10). The center N region is called the base. This base is usually 0.001 inch thick. A collector junction and an emitter junction are also formed. Note the emitter is represented by an arrow either going toward the vertical line or away from it. The collector is the line without an arrow. The third line, the one that has the other two contacting it, is the base. These are usually abbreviated C, B, and E.

Transistors can be used for amplification of signals or they can be used for switching. There are three common configurations for transistor circuits. They are the *common base, common collector,* and *common emitter.* The common-emitter circuit is the most often used (see Fig. 11-11). In this type of circuit the current through the load flows between the emitter and collector. The input signal is applied between the emitter and the base. In normal operation, the collector junction is reverse biased by the supply voltage, B_1. The emitter junction is forward biased by the applied voltage, B_2. Electrons flow across the forward-biased emitter into the base. They diffuse through the base region and flow across the collector junction. Then they flow through the external collector circuit.

Battery B_2 voltage is applied in the forward direction. This means the voltage is positive to the emitter P-type crystal. It also means that voltage is negative to the N-type crystal. Thus, the emitter-base junction has a low impedance.

The voltage of battery B_1 is applied in the reverse direction. This means the

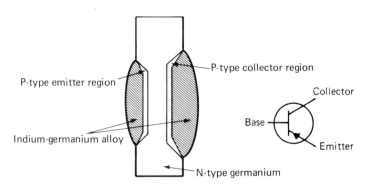

FIGURE 11-10 Transistor junction using germanium and indium.

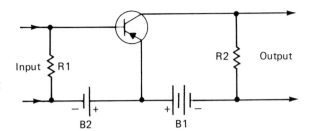

FIGURE 11-11 Schematic diagram of a common-emitter circuit using a PNP transistor.

voltage is positive to the N-type crystal. It also means that voltage is negative to the P-type crystal. This, then, produces a collector-base junction with a high impedance.

Transistor Impedances

The impedance of the emitter junction is low. Thus, electrons flow from the emitter region to the base region. At the junction, the electrons combine with the holes in the N-type base crystal. If the base is thin enough, almost all the holes are attracted to the negative terminal of the collector. They then flow through the load to B_1.

The collector current is stopped by applying a positive voltage to the base and a negative voltage to the emitter. In actual transistors, however, this cannot be done because of several basic limitations. Some of the electrons in the base region flow across the emitter junction. Some combine with the holes in the base region. For this reason, it is necessary to supply a current to the base. This makes up for these losses.

The ratio of the collector current to the base current is known as the *current gain* of the transistor. Current gain, called *beta* (β), is found by dividing base current into collector current. At high frequencies, the fundamental limitation is the time for carriers to diffuse across the base region. They move from the emitter to the collector. This is why the base region width or thickness is so important. The thinner the base region, the less time is required for the carriers to diffuse across it. This causes the transistor to operate faster.

Figure 11-12 shows a schematic for a common-base PNP circuit. The signal is injected into the emitter-base circuit. The output signal is taken from the collec-

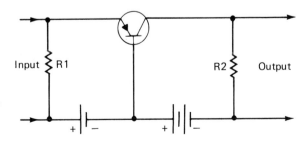

FIGURE 11-12 Common-base circuit using a PNP transistor.

tor-base circuit. An important advantage of the transistor is its ability to transfer impedances. This is where the transistor gets its name. The word transistor comes from *trans*fer res*istor.*

The emitter circuit has low impedance. This low impedance allows current to flow. This current flow then creates a current through the collector circuit. The emitter has low impedance and low current. The collector has high impedance and even slightly less current than the emitter. However, more power is the result in the collector. This is because $P = I^2R$ or, in this case, $P = I^2Z$. Impedance (Z) is high and the current is squared. Thus, the collector circuit has more power than the emitter circuit with its low impedance.

A common-emitter circuit has about 1300 ohms of input impedance. This is compared to an output impedance of about 50,000 ohms. Thus, there is an increase in impedance from emitter to collector of about 39 times. The junction transistor amplifies in this way. It acts as a power amplifier. A small change in emitter voltage causes a large change in the collector circuit. The different impedances cause this reaction.

The power transistor is shown in Fig. 11-13. Along with the power transistor is the much smaller signal amplifying and switching transistor normally seen in electronics equipment.

INTEGRATED CIRCUITS

The semiconductor monolithic chip was developed in 1958 by J. S. Kilby. Active and passive circuit components were successively diffused and deposited on a single chip. Shortly after, Robert Noyce made a complete circuit on a single chip. This was

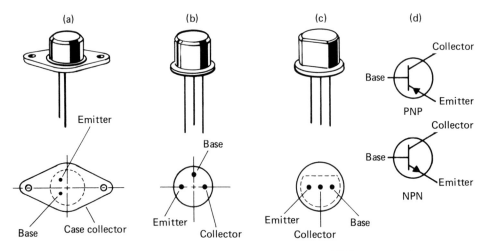

FIGURE 11-13 Three types of transistor packages. (A) Power transistors. (B) TO-5 case. (C) Small-signal transistors. (D) Transistor symbols.

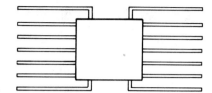
FIGURE 11-14 Flat pack IC.

the beginning of the modern, inexpensive integrated circuit (IC or chip). Resistors, capacitors, transistors, and diodes can be placed on a chip. These chips are available in three standard packages, as follows.

Flat Pack. The flat pack is hermetically sealed. This means it is vacuum packed. The ceramic flat pack has either 10 or 14 pins. This type of packaging is no longer used. It may be seen in some older-model equipment with chips (see Fig. 11-14).

Multipin Circular. This type of packaging is no longer in common use. It was originally used because it fit into the same type of container as a transistor. It had more than three leads coming from a TO-5 case originally made for transistors (see Fig. 11-15).

Dual In-line Package (DIP). This is the most commonly used type of integrated circuit today. When you think of a "chip" you think of this type. It may have many leads from each side (see Fig. 11-16). This is an easily used type of IC since it can be placed into a socket and removed if it develops any problem. This type of package is standardized as to size. It can be placed into circuit boards by machines, which increases its usefulness in electronics equipment. Many thousands of ICs can be made at a time. This means the cost of manufacturing is very low. DIPs are used in computers, calculators, and control devices for air conditioning and heating equipment. Amplifiers are fabricated as complete units. Everything seems to use ICs today. Toys, calculators, computers, and automobiles all use ICs or chips in one form or another. The possibilities are unlimited. ICs will play a more important role in air-conditioning, refrigeration, and heating devices in the future.

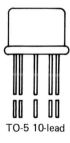
TO-5 10-lead

FIGURE 11-15 TO-5 case IC.

FIGURE 11-16 Integrated circuits in dual-in-line packages (DIPs).

Solid-State Demand Defrost Control

An excellent example of a refrigeration control using semiconductor technology is the *demand defrost control.* The idea of defrosting only when needed saves money and energy and protects the quality of the refrigerated product, since extremes in temperature swings are minimized.

THERMISTOR SENSING

A temperature difference concept can be used to automatically initiate the defrost cycle on vertical open frozen food display cases that use electric defrost (see Fig. 11-17). If the cover is lifted, you can see the electronic package. There are a few transistors and the usual capacitors, resistors, and a transformer. The sensor that feeds in the information needed to operate properly as a control is shown in Fig. 11-18. These two sensors are simple in construction. They are simple *thermistors* that change resistance when temperature changes (see Fig. 11-19). Their operation

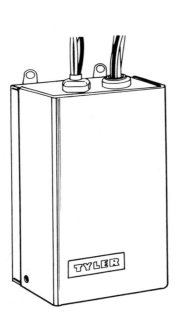

FIGURE 11-17 DF-1 solid-state demand defrost control for vertical open frozen food display cases. (Courtesy of Tyler)

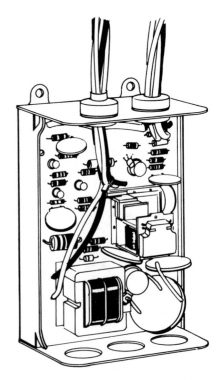

FIGURE 11-18 Cover removed from the DF-1 defrost control. (Courtesy of Tyler)

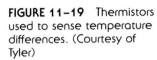
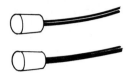

FIGURE 11-19 Thermistors used to sense temperature differences. (Courtesy of Tyler)

is not affected by dirt, moisture, lint, food particles, or ice. They can sense the least temperature change precisely.

Figure 11-20 shows how frost buildup is sensed to cause the circuit to know what is happening. This tells when defrost should be started and terminated. The ever widening temperature difference is a sure sign of frost buildup. The temperature difference of air flowing over a coil increases in direct proportion to frost buildup. This is what the thermistor monitors to trigger defrost only when needed. Refrigeration fixtures run at peak efficiency all the time. They use less energy and keep the product at a lower, steadier temperature.

HUMIDITY SENSING

Figure 11-21 shows how defrost frequency and humidity are related. Improvements in the design of humidity-sensing elements and the materials used in their construction have minimized many past limitations of humidity sensors. One type of humidity sensor used with electronic controls is a resistance CAB (cellulose acetate butyrate) element (see Fig. 11-22). This resistance element is an improvement over other resistance elements. It has greater contamination resistance, stability, and durability. The humidity CAB element is a multilayered, humidity-sensitive, polymeric film. It

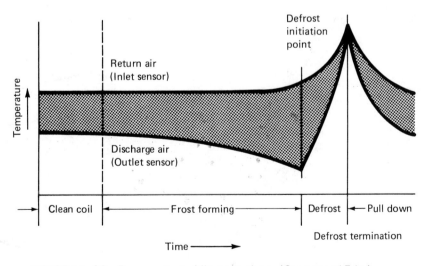

FIGURE 11-20 Temperature-difference chart. (Courtesy of Tyler)

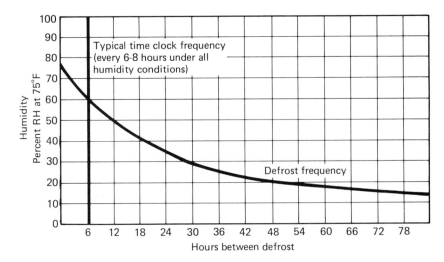

FIGURE 11-21 Effect of humidity on hours between defrosting. (Courtesy of Tyler)

consists of an electrically conductive core and insulating outer layers. These layers are partially hydrolyzed. The element has a nominal resistance of 2500 ohms. It has a sensitivity of 2 ohms per 1% relative humidity (rh) at 50% rh. Its humidity sensing range is rated at 0% to 100% rh.

The CAB element consists of conductive humidity-sensitive film, mounting components, and a protective cover (see Fig. 11-23). The principal component of this humidity sensor is the film. The film has five layers of CAB in the form of a ribbon strip. The CAB material is used because of its good chemical and mechanical stability and high sensitivity to humidity. It also has excellent film-forming characteristics.

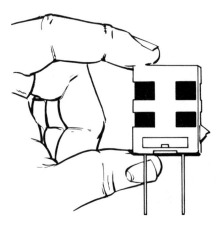

FIGURE 11-22 CAB resistive element.

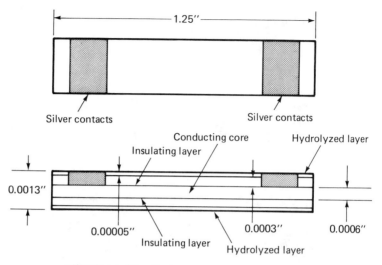

FIGURE 11-23 Hydrolyzed humidity element.

The CAB resistance element is a carbon element having a resistance/humidity tolerance favorable to inclusion in a control circuit. With an increase in relative humidity, water is absorbed by the CAB, causing it to swell. This swelling of the polymer matrix causes the suspended carbon particles to move farther apart from each other. This results in an increased element resistance.

When relative humidity decreases, water is given up by the CAB. The contraction of the polymer causes the carbon particles to come closer together. This, in turn, makes the element more conductive or less resistive.

Bridge Circuit

A bridge circuit is a network of resistances and capacitive or inductive impedances. The bridge circuit is usually used to make precise measurements. The most common bridge circuit is the Wheatstone bridge. This consists of variable and fixed resistances. Simply, it is a series–parallel circuit (see Fig. 11-24). The branches of the circuit forming the diamond shape are called legs.

If 10 volts dc is applied to the bridge circuit shown in Fig. 11-25, one current will flow through R_1 and R_2 and another through R_3 and R_4. Since R_1 and R_2 are both fixed 1000-ohm resistors, the current through them is constant. Each resistor will drop one-half of the battery voltage, or 5 volts. Thus, 5 volts is dropped across each resistor. The meter senses the sum of the voltage drops across R_2 and R_3. Both are 5 volts. However, the R_2 voltage drop is a positive (+) to negative (−) drop. The R_3 drop is a negative to positive drop. They are opposite in polarity and cancel each other. This is called a *balanced bridge*. The actual resistance values are not important. What is important is that this ratio be maintained and the bridge be balanced.

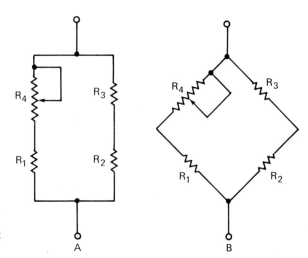

FIGURE 11-24 Bridge-circuit configurations.

Unbalanced Bridge

In Fig. 11-26, the value of the variable resistor R_4 is 950 ohms. The other resistors have the same value. Using Ohm's law, the voltage drop across R_4 is found to be 4.9 volts. The remaining voltage, 5.1 volts, is dropped across R_3 (see Fig. 11-26). The voltmeter measures the sum of the voltage drops across R_2 and R_3 as 5.0 volts (+ to −) and 5.1 volts (− to +). It registers a total of −0.1 volt.

In Fig. 11-27, the converse is true. The value of R_4 is 1050 ohms. The voltage

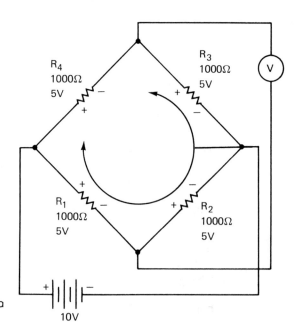

FIGURE 11-25 Current in a balanced bridge circuit.

HUMIDITY SENSING 207

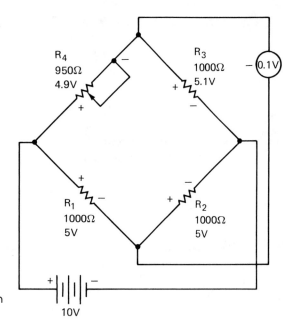

FIGURE 11-26 Operation of a bridge circuit.

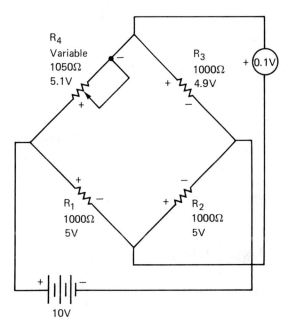

FIGURE 11-27 Operation of a bridge circuit.

drop across R_3 is 4.9 volts. The voltmeter senses the sum of 5 volts (+ to −) and 4.9 volts (− to +), or +0.1 volt.

When R_4 changes the same amount above or below the balanced-bridge resistance, the magnitude of the dc output measured by the voltmeter is the same. However, the polarity is reversed.

Sensors

The *sensor* in a control system is a resistance element that varies in resistance value with changes in the variable it is measuring. This may be humidity or temperature. These resistance changes are converted into proportional amounts of voltage by a bridge circuit. The voltage is amplified and used to position actuators that regulate the controlled variable.

CONTROLLERS

The *sensing bridge* is the section of the controller circuit that contains the temperature-sensitive element or elements. The potentiometer for establishing the set point is also part of the control system. The bridges are energized with a dc voltage. This permits long wire runs in sensing circuits without the need for compensating wires or for other capacitive compensating arrangements.

Both integral (room) and remote-sensing element controllers produce a proportional 0- to 16-volt dc output signal in response to a measured temperature change. Controllers can be wired to provide direct or reverse action. Direct-acting operation provides an increasing output signal in response to an increase in temperature. Reverse-acting operation provides an increasing output signal in response to a decrease in temperature.

ELECTRONIC CONTROLLERS

Electronic controllers have three basic elements: the bridge, the amplifier, and the output circuit. Two legs of the bridge are variable resistances in Fig. 11–28. The sensor and the set-point potentiometer are shown in the bridge configuration. If temperature changes or if the set point is changed, the bridge is in an *unbalanced state*. This gives a corresponding output result. The output signal, however, lacks power to position actuators. Therefore, this signal has to be amplified to become useful in the control of devices associated with making sure the right amount of heat or cooled air gets to the room intended.

DIFFERENTIAL AMPLIFIERS

Controllers utilize direct-coupled dc differential amplifiers to increase the millivolt signal from the bridge to the necessary 0- to 16-volt level for the actuators. There

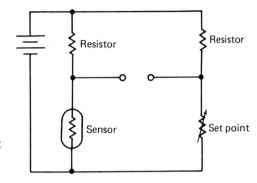

Figure 11-28 Bridge circuit with a sensor and a set point.

are two amplifiers, one for direct reading and the other for reversing signals. Each amplifier has two stages of amplification. This arrangement is shown in block form in Fig. 11-29.

The differential transistor circuit provides gain and good temperature stability. Figure 11-30 compares a single transistor amplifier stage with a differential amplifier. Transistors are temperature sensitive. That is, the currents they allow to pass depend on the voltage at the transistor and its ambient temperature. An increase in the ambient temperature in the circuit shown in Fig. 11-30A causes the current through the transistor to increase. The output voltage decreases. The emitter resistor, R_E, reduces this temperature effect. It also reduces the available voltage gain in the circuit because the signal voltage across the resistor amounts to a negative feed-

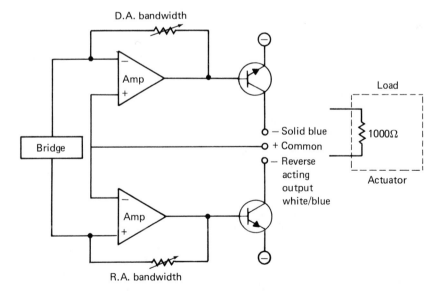

Figure 11-29 DC differential amplifier for use in a controller circuit.

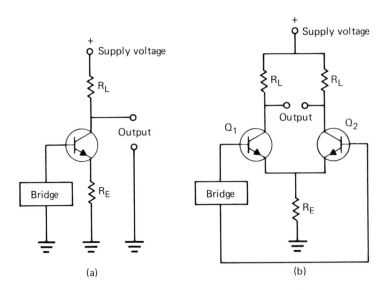

Figure 11-30 (A) Single transistor amplifier stage. (B) Two-transistor amplifier stage.

back voltage. That is, it causes a decrease in the voltage *difference* that was originally produced by the change in temperature at the sensing element.

Since it is desirable for the output voltage of the controller to correspond only to the temperature of the sensing elements and not to the ambient temperature of the amplifier, the circuit shown in Fig. 11-30B is used. Here any ambient temperature changes affect both transistors at the same time. The useful output is taken as the difference in output levels of each transistor, and the effects of temperature changes are canceled. The voltage gain of the circuit shown in Fig. 11-30B is much higher than that shown in Fig. 11-30A. This is because the current variations in the two transistors produced by the bridge signal are equal and opposite. An increase in current through Q_1 is accompanied by a decrease in current through Q_2. The sum of these currents through R_E is constant. No signal voltage appears at the emitters to cause negative feedback as in Fig. 11-30A.

The result of sequentially varying dc signals in response to temperature change at the sensing element is shown in Fig. 11-31.

ACTUATORS

Cybertronic actuators perform the work in an electronic system. They accept a control signal and translate that signal into mechanical movement. This is used to position valves and dampers. The electrohydraulic actuators are so called because they convert electrical signals into a fluid movement or force. Damper actuators,

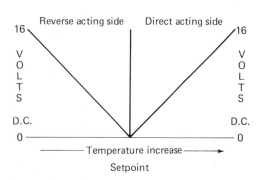

Figure 11-31 Result of sequentially varying dc signals in response to temperature change at the sensing element.

equipped with linkage for connection to dampers, and value actuators, having a yoke and linkage to facilitate mounting on a valve body, are also available.

OTHER DEVICES

Low-and high-signal selectors accept several control signals. Such selectors then compare the signals and pass the lowest or highest. For example, a high-signal selector can be used on a multizone unit to control the cooling coil. The zone requiring the most cooling transmits the highest control signal. This, in turn, will be passed by the high-signal selector to energize the cooling.

The future of electronics in the control of heating, cooling, and refrigeration is unlimited. Only a few examples have been given. More and more mechanically operated thermostats and valves will be replaced and become electronically controlled. A computer has already been utilized in making a programmable controller that can process the information fed to it by many inputs. This will improve process control in industry and increase the comfort of large buildings by more accurately controlling the flow of heat or cooled air. The computer also aids in locating the source of trouble in a system, thereby eliminating a lot of troubleshooting.

SOLID-STATE COMPRESSOR MOTOR PROTECTION

Solid-state circuitry for air-conditioning units has been in use for some time. The following is an illustration of how some of the circuitry has been incorporated into the protection of compressor motors. This module is manufactured by Robertshaw Controls Co. of Milford, Connecticut.

Solid-state motor protection prevents motor damage caused by excessive temperature in the stator windings. These solid-state devices provide excellent phase-leg protection by means of separate sensors for each phase winding. The principal advantage of this solid-state system is its speed and sensitivity to motor temperature and its automatic reset provision.

There are two major components to the protection system:

1. The protector sensors are embedded in the motor windings at the time the motor is manufactured.
2. The control module is a sealed enclosure containing a transformer and a switch. Figure 11-32 shows two models.

Operation

Leads from the internal motor sensors are connected to the compressor terminals as shown in Fig. 11-33. Leads from the compressor terminals to the control module are connected as shown in Fig. 11-34. Figure 11-34A shows the older model and Fig. 11-34B the newer model. While the exact internal circuitry is quite complicated, basically the modules sense resistance change through the sensors as the result of motor temperature changes in the motor windings. This resistance change triggers the action of the control circuit relay at predetermined opening and closing settings, which causes the line voltage circuit to the compressor to be broken and completed, respectively.

The modules are available for either 208/240- or 120-volt circuits. The module is plainly marked as to the input voltage. The sensors operate at any of the stated voltages because an internal transformer provides the proper power for the solid-state components.

The two terminals on the module marked Power Supply (T1 and T2) are con-

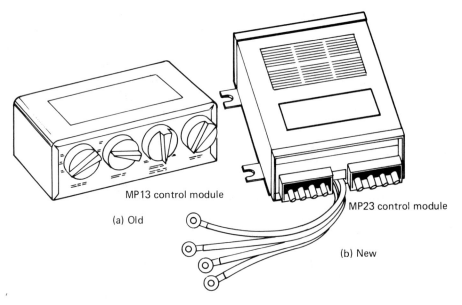

Figure 11-32 Solid-state control modules. (A) Older unit. (B) Newer unit. (Courtesy of Robertshaw)

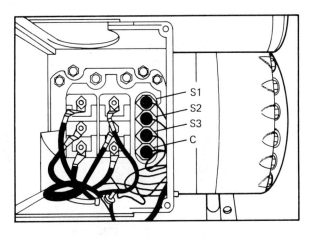

Figure 11-33 Compressor terminal board. (Courtesy of Robertshaw)

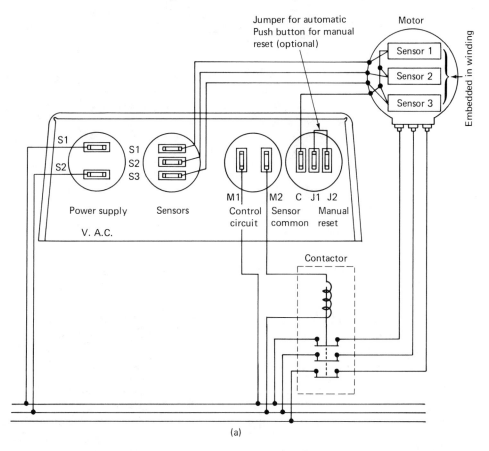

(a)

Figure 11-34 Solid-state control modules. (A) Older unit wiring details. (B) Newer unit wiring details.

214 SOLID-STATE CONTROLS

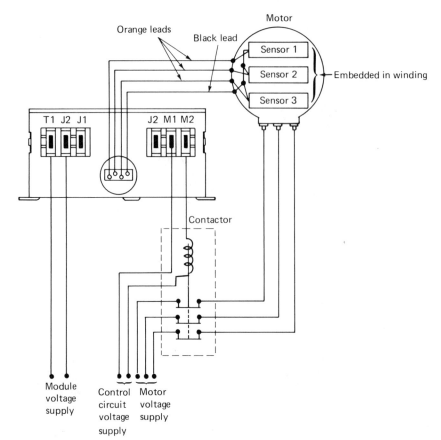

Note: Control is automatic reset when terminals J1 and J2 are not included. The control is manual reset when terminals J1 and J2 are included.

(b)

Figure 11-34 (B) (*continued*)

nected to a power source of the proper voltage, normally the line terminals on the compressor motor contactor or the control circuit transformer as required.

Troubleshooting the Control

The solid-state module cannot be repaired in the field, and if the cover is opened or the module physically damaged, the warranty on the module is voided. No attempt should be made to adjust or repair this module, and if it becomes defective, it must be returned intact for replacement. This is the usual procedure for most solid-state units. However, if the unit becomes defective, you should be able to recognize that fact and replace it.

If the compressor motor is inoperable or is not operating properly, the solid-state control circuit may be checked as follows:

1. If the compressor has been operating and has tripped on the protector, allow the compressor to cool for at least 1 hour before checking to allow time for the motor to cool and the control circuit to reset.
2. Connect a jumper wire across the control circuit terminals on the terminal board (see Fig. 11-34). This will bypass the relay in the module. If the compressor will not operate with the jumper installed, then the problem is external to the solid-state protection system. If the compressor operates with the module bypassed, but will not operate when the jumper wire is removed, then the control circuit relay is open.
3. If, after allowing time for motor cooling, the protector still remains open, the motor sensors may be checked as follows:
    a. Remove the wiring connections from the sensor and common terminals on the compressor board (see Figs. 11-33 and 11-34).
    b. **WARNING:** Use an ohmmeter with a 3-volt maximum battery power supply. The sensors are sensitive and easily damaged, and no attempt should be made to check continuity through them. Any external voltage or current applied to the sensors may cause damage, necessitating compressor replacement.
    c. Measure the resistance from each sensor terminal to the common terminal. The resistance should be in the following range: 75 ohms (cold) to 125 ohms (hot). Resistance readings in this range indicate the sensors are good. A resistance approaching zero indicates a short. A resistance approaching infinity indicates an open connection. If the sensors are damaged, they cannot be repaired or replaced in the field, and the compressor must be replaced to restore motor protection.
4. If the sensors have proper resistance and the compressor will run with the control circuit bypassed but will not run when connected properly, the solid-state module is defective and must be replaced. The replacement module must be the same voltage and made by the same manufacturer as the original module on the compressor.

Restoring Service

In the unlikely event that one sensor is damaged and has an open circuit, the control module will prevent compressor operation even though the motor may be in perfect condition. If such a situation should be encountered in the field, as an *emergency* means of operating the compressor until such time as a replacement can be made, a properly sized resistor can be added between the terminal of the open sensor and the common sensor terminal in the compressor terminal box (see Figs. 11-33 and 11-35). This then indicates to the control module an acceptable resistance in the

216 SOLID-STATE CONTROLS

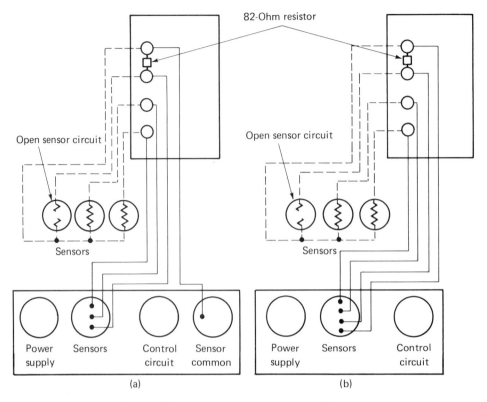

Figure 11-35 Adding a resistor to compensate for an open sensor. (Courtesy of Robertshaw)

damaged sensor circuit, and compressor operation can be restored. The emergency resistor should be a 2 watt, 82 ohm, wire wound with a tolerance of ±5%.

In effect, the compressor will continue operation with two-leg protection rather than three-leg protection. While this obviously does not provide the same high degree of protection, it does provide a means of continuing compressor operation with a reasonable degree of safety.

REVIEW QUESTIONS

1. What does the word semiconductor mean?
2. What are the two materials used for semiconductor devices?
3. What is a diode? What is the PN junction diode used for?
4. What are the uses for diodes?
5. What is an SCR? Where is it used?
6. What are the two main uses for transistors?

7. What is a transistor's current gain?
8. What is an integrated circuit?
9. What is a thermistor?
10. What does CAB stand for in a humidity element?
11. What is a bridge circuit?
12. How do balanced and unbalanced bridge circuits differ?
13. What is a sensor?
14. How is a sensing bridge connected?
15. What is an actuator?
16. What is a differential amplifier used for?

PERFORMANCE OBJECTIVES

Know how an induction motor works.
Know how to identify a shaded-pole motor and where it is used.
Know how to identify a split-phase motor and where it is used.
Know how to identify a three-phase motor and where it is used.
Know how a capacitor-start motor works.
Know how a permanent split-capacitor motor works.
Know where a capacitor-start, capacitor-run motor is used.
Know how a three-phase motor is electrically connected and properly maintained.

CHAPTER 12

Alternating Current Motors

Most ac motors are of the induction type. They are, in general, simpler and cheaper to build than equivalent dc machines. They have no commutator, slip rings, or brushes, and there is no electrical connection to the rotors. Only the stator winding is connected to the ac source, and, then, as their name implies, induction produces the currents in the rotor. A common and particularly simple form of rotor for this type of motor is the squirrel-cage rotor (see Fig. 12-1). It is so named because of its resemblance to a treadmill-type squirrel cage. The induction motor is based on a rotating magnetic field. This is achieved by using multiple stator field windings (poles), each pair of which is excited by an ac voltage of the same amplitude and frequency as, but phase-displaced from, the voltage supplying the neighboring pair. Figure 12-2 shows how the magnetic field rotates in a four-pole induction motor,

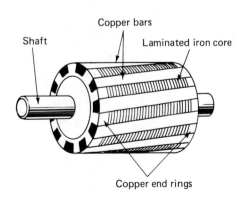

FIGURE 12-1 Squirrel-cage rotor.

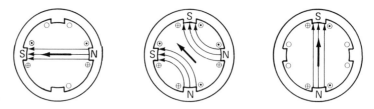

FIGURE 12-2 Rotating field for a four-pole stator.

where the voltages to the two pairs of poles are 90° out of phase with each other. When the rotor is placed in the stator's rotating field, the induced currents set up their own fields, which react with the stator's field and push the rotor around.

Note that some rotors are *skewed*. The skew of a rotor refers to the amount of angle between the conductor slots and the end face of the rotor laminations. Normally, the conductors are in a nearly straight line, but for high torque applications the rotor is skewed, which increases the angle of the conductors. The term *full skew* refers to the maximum practical amount (see Fig. 12-3). Figure 12-4 shows how the rotor is located in reference to the stator and the end bells that hold it in place.

A number of types of ac motors are available. The types presented here are those most often encountered when working with heating, air-conditioning, and refrigeration equipment.

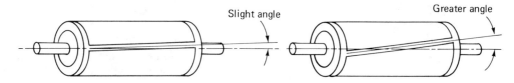

FIGURE 12-3 Skewed die-cast rotor. Note the angle of the conductor slots.

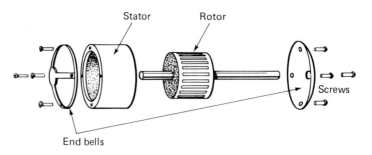

FIGURE 12-4 Exploded view of motor.

SHADED-POLE MOTOR

The shaded-pole induction motor is a single-phase motor. It uses a unique method to start the rotor turning. The effect of a moving magnetic field is produced by constructing the stator in a special way (see Fig. 12-5).

Portions of the pole piece surfaces are surrounded by a copper strap called a shading coil. The strap causes the field to move back and forth across the face of the pole piece. In Fig. 12-6, a numbered sequence and points on the magnetization curve are shown. As the alternating stator field starts increasing from zero (1), the lines of force expand across the face of the pole piece and cut through the strap. A voltage is induced in the strap. The current that results generates a field that opposes the cutting action (and decreases the strength) of the main field. This action causes certain actions: As the field increases from zero to a maximum of 90°, a large portion of the magnetic lines of force is concentrated in the unshaded portion of the pole (1). At 90° the field reaches its maximum value. Since the lines of force have stopped expanding, no emf is induced in the strap, and no opposite magnetic field is generated. As a result, the main field is uniformly distributed across the poles as shown in (2).

From 90° to 180°, the main field starts decreasing or collapsing inward. The field generated in the strap opposes the collapsing field. The effect is to concentrate the lines of force in the shaded portion of the poles, as shown in (3).

Note that from 0° to 180° the main field has shifted across the pole face from the unshaded to the shaded portion. From 180° to 360°, the main field goes through the same change as it did from 0° to 180°. However, it is now in the opposite direction (4). The direction of the field does not affect the way the shaded pole works.

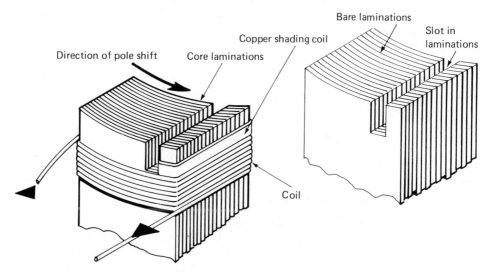

FIGURE 12-5 Shading of the poles of a shaded-pole motor.

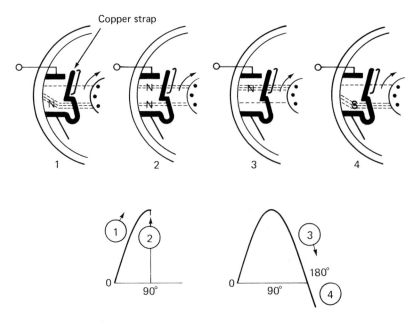

FIGURE 12-6 Shaded poles as used in shaded-pole ac motors.

The motion of the field is the same during the second half-hertz as it was during the first half-hertz.

The motion of the field back and forth between shaded and unshaded portions produces a weak torque. This torque is used to start the motor. Because of the weak starting torque, shaded-pole motors are built in only small sizes. They drive such devices as fans, timers, and blowers.

Reversibility. Shaded-pole motors can be reversed mechanically. Turn the stator housing and shaded poles end for end. These motors are available from 1/25 to 1/2 horsepower.

Uses. As previously mentioned, this type of motor is used as a fan motor in refrigerators and freezers and in some types of air-conditioning equipment where the demand is not too great. It can also be used as part of a timing device for defrost timers and other sequenced operations.

The fan and motor assembly are located behind the provisions compartment in the refrigerator, directly above the evaporator in the freezer compartment. The suction type fan pulls air through the evaporator and blows it through the provisions compartment air duct and freezer compartment fan grille. Figure 12-7 shows a shaded-pole motor with a molded plastic fan blade. For maximum air circulation, the location of the fan on the motor shaft is most important. Mounting the fan blade too far back or too far forward on the motor shaft, in relation to the evapora-

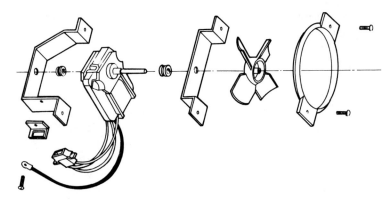

FIGURE 12-7 Fan, motor, and bracket assembly. (Courtesy of Kelvinator)

tor cover, will result in improper air circulation. The freezer compartment fan must be positioned with the lead edge of the fan 1/4 inch in front of the evaporator cover.

The fan assembly shown in Fig. 12-8 is used on top freezer, no-frost, fiberglass-insulated model refrigerators. The freezer fan and motor assembly is located in the divider partition directly under the freezer air duct.

SPLIT-PHASE MOTOR

The field of a single-phase motor, instead of rotating, merely pulsates. No rotation of the rotor takes place. A single-phase pulsating field may be visualized as two rotating fields revolving at the same speed, but in opposite directions. It follows,

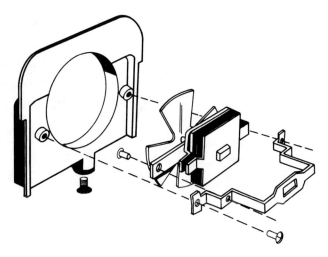

FIGURE 12-8 Fan and fan motor bracket assembly. (Courtesy of Kelvinator)

therefore, that the rotor will revolve in either direction at nearly synchronous speed, if it is given an initial impetus in either one direction or the other. The exact value of this initial rotational velocity varies widely with different machines. A velocity higher than 15% of the synchronous speed is usually sufficient to cause the rotor to accelerate to the rated or running speed. A single-phase motor can be made self-starting if means can be provided to give the effect of a rotating field.

To get the split-phase motor running, a *run winding* and a *start winding* are incorporated into the stator of the motor. Figure 12-9 shows the split-phase motor with the end cap removed so you can see the starting switch and governor mechanism.

This type of motor is difficult to use with air conditioning and refrigeration equipment inasmuch as it has very little starting torque and will not be able to start a compressor since it presents a load to the motor immediately upon starting. This type of motor, however, is very useful in heating equipment (see Fig. 12-10).

Getting the Motor Started

One of the most important parts of the single-phase electric motor is the start mechanism. A special type is needed for use with single-phase motors. A centrifugal switch is used to take a start winding out of the circuit once the motor has come up to within 75% of its run speed. The split-phase, capacitor-start, and other variations of these types all need the start mechanism to get them running.

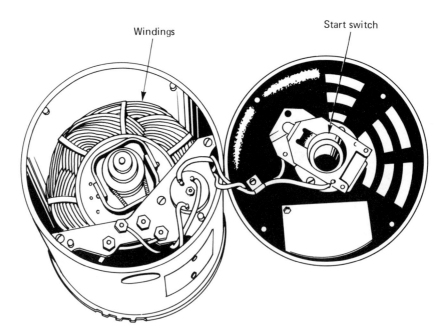

FIGURE 12-9 Single-phase starting switch and governor mechanism.

SPLIT-PHASE MOTOR 225

FIGURE 12-10 Single-phase, split-phase furnace motor.

The stator of a split-phase motor has two types of coils; one is called the run winding and the other the start winding. The run winding is made by winding the enamel-coated copper wire through the slots in the stator punchings.

The *start winding* is made in the same way except that the wire is smaller. Coils that form the start windings are positioned in pairs in the stator directly opposite each other and between the run windings. When you look at the end of the stator, you see alternating run windings and start windings (see Fig. 12-9).

The run windings are all connected together, so the electrical current must pass through one coil completely before it enters the next coil, and so on through all the run windings in the stator. The start windings are connected together in the same way, and the current must pass through each in turn (see Fig. 12-11).

The two wires from the run windings in the stator are connected to terminals on an insulated terminal block in one end bell where the power cord is attached to the same terminals. One wire from the start winding is tied to one of these terminals also. However, the other wire from the start winding is connected to the stationary switch mounted in the end bell. Another wire then connects this switch to the opposite terminal on the insulated block. The stationary switch does not revolve, but is placed so the weights in the rotating portion of the switch, located on the rotor, will move outward when the motor is up to speed and open the switch to stop electrical current from passing through the start winding.

The motor then runs only on the main winding until such time as it is shut off. Then, as the rotor decreases in speed, the weights on the rotating switch again move inward to close the stationary switch and engage the start winding for the next time it is started.

Reversibility. The direction of rotation of the split-phase motor can be changed by reversing the start winding leads.

Uses. This type of motor is used for fans, furnace blowers, oil burners, office appliances, and unit heaters.

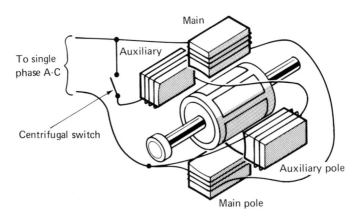

FIGURE 12-11 Single-phase induction motor.

REPULSION START, INDUCTION RUN MOTOR

The repulsion-induction motor starts on one principle of operation and, when almost up to speed, changes over to another type of operation. Very high twisting forces are produced during starting by the repulsion between the magnetic pole in the armature and the same kind of pole in the adjacent stator field winding. The repulsing force is controlled and changed so that the armature rotational speed increases rapidly, and, if not stopped, would continue to increase beyond a practical operating speed. It is prevented by a speed-actuated mechanical switch that causes the armature to act as a rotor that is electrically the same as the rotor in single-phase induction motors. That is why the motor is called a repulsion-induction motor.

The stator of this motor is constructed very much like that of a split-phase or capacitor-start motor, but only run or field windings are mounted inside. End bells keep the armature and shaft in position and hold the shaft bearings.

The armature consists of many separate coils of wire connected to segments of the commutator. Mounted on the other end of the armature are governor weights that move push rods that pass through the armature core. These rods push against a short-circuiting ring mounted on the shaft on the commutator end of the armature. Brush holders and brushes are mounted in the commutator end bell, and the brushes, connected by a heavy wire, press against segments on opposite sides of the commutator (see Fig. 12-12).

When the motor is stopped, the action of the governor weights keeps the short-circuiting ring from touching the commutator. When the power is turned on and current flows through the stator field windings, a current is induced in the armature coils. The two brushes connected together form an electromagnetic coil that produces a north and south pole in the armature, positioned so that the north pole in the armature is next to a north pole in the stator field windings. Since like poles try to move apart, the repulsion produced in this case can be satisfied in only one way: the armature turns and moves the armature coil away from the field windings.

The armature turns faster and faster, accelerating until it reaches what is approximately 80% of the run speed. At this speed, the governor weights fly outward and allow the push rods to move. These push rods, which are parallel to the armature shaft, have been holding the short-circuiting ring away from the commutator. Now that the governor has reached its designed speed, the rods can move together electrically in the same manner that the cast aluminum discs did in the cage of the induction motor rotor. This means that the motor runs as an induction motor.

The repulsion-induction type of motor can start very heavy, hard-to-turn loads without drawing too much current. They are made from 1/2 to 20 horsepower. This type motor is used for such applications as large air compressors, refrigeration equipment, and large hoists and is particularly useful in locations where low line voltage is a problem.

This type of motor is no longer used in the refrigeration industry. Some older operating units may be found with this type of motor still in use.

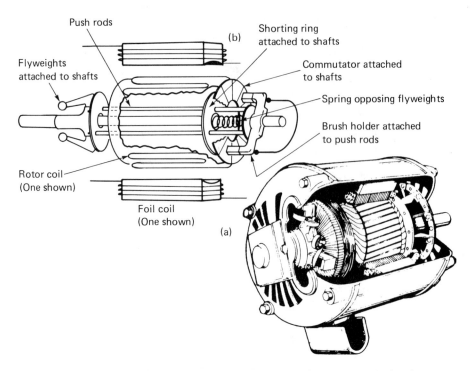

FIGURE 12-12 A. Brush-lifting, repulsion-start, induction-run, single-phase motor. B. Brush-lifting, repulsion-start, induction-run motor.

CAPACITOR-START MOTOR

The capacitor motor is slightly different from a split-phase motor. A capacitor is placed in the path of the electrical current in the start winding (see Fig. 12-13). Except for the capacitor, which is an electrical component that slows any rapid change in current, the two motors are the same electrically. A capacitor motor can usually be recognized by the capacitor can or housing that is mounted on the stator (see Fig. 12-14).

Adding the capacitor to the start winding increases the effect of the two-phase field described in connection with the split-phase motor. The capacitor means that the motor can produce a much greater twisting force when it is started. It also reduces the amount of electrical current required during starting to about 1.5 times the current required after the motor is up to speed. Split-phase motors require three or four times the current in starting that they do in running.

Reversibility. An induction motor will not always reverse while running. It may continue to run in the same direction, but at a reduced efficiency. An inertia-type load is difficult to reverse. Most motors that are classified as reversible while run-

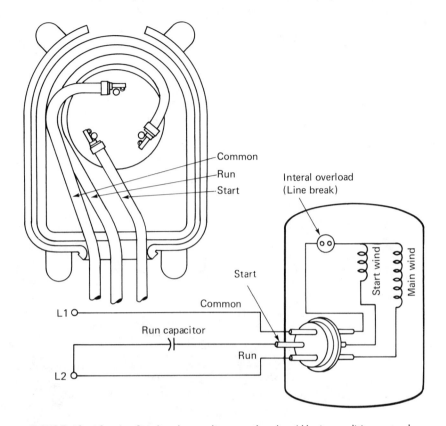

FIGURE 12-13 A. Single phase diagram for the AH air conditioner and heat-pump compressor. (Courtesy of Tecumseh) B. Terminal box showing the position of the terminals on the AH series of compressors. (Courtesy of Tecumseh)

ning will reverse with a noninertial-type load. They may not reverse if they are under no-load conditions or have a light load or an inertial load.

One problem related to the reversing of a motor while it is still running is the damage done to the transmission system connected to the load. In some cases it is possible to damage a load. One way to avoid this is to make sure the right motor is connected to a load.

Reversing (while standing still) the capacitor-start motor can be done by reversing its start winding connections. This is usually the only time that a field technician will work on a motor. The available replacement motor may not be rotating in the direction desired, so the technician will have to locate the start winding terminals and reverse them in order to have the motor start in the desired direction.

Uses. Capacitor motors are available in sizes from 1/6 to 20 horsepower. They are used for fairly hard starting loads that can be brought up to run speed in under 3

FIGURE 12-14 Capacitor-start motor.

seconds. They may be used in industrial machine tools, pumps, air conditioners, air compressors, conveyors and hoists.

Figure 12-15 shows a capacitor-start, induction-run motor used in a compressor. This type uses a relay to place the capacitor in and out of the circuit. More about this type of relay will be discussed later. Figure 12-16 shows how the capacitor is located outside the compressor.

CAPACITOR-START MOTOR

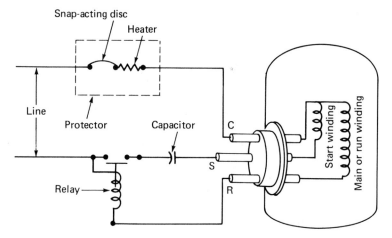

FIGURE 12-15 Capacitor-start, induction-run motor used for a compressor. (Courtesy of Tecumseh)

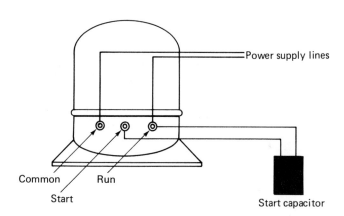

Start capacitor* sizes	
Compressor 1/8 hp	Capacitor size is 95 to 200 μF
Compressor 1/6 hp	Capacitor size is 95 to 200 μF
Compressor 1/4 hp	Capacitor size is 200 to 300 μF
Compressor 1/3 hp	Capacitor size is 250 to 350 μF
Compressor 1/2 hp	Capacitor size is 300 to 400 μF
Compressor 3/4 hp	Capacitor size is 300 to 400 μF

Black case (Bakelite)

FIGURE 12-16 Location of start capacitor in a compressor circuit. (Courtesy of Tecumseh)

232 ALTERNATING CURRENT MOTORS

PERMANENT SPLIT-CAPACITOR (PSC)

The permanent split-capacitor motor is used in compressors for air-conditioning and refrigeration units. It has an advantage over the capacitor-start motor inasmuch as it does not need a centrifugal switch with its associated problems.

The PSC motor has a run capacitor in series with the start winding. Both run capacitor and start winding remain in the circuit during start and after the motor is up to speed. Motor torque is sufficient for capillary and other self-equalizing systems. No start capacitor or relay is necessary. The PSC motor is basically an air-conditioner compressor motor. It is also used in refrigerator compressors. It is very common through 3 horsepower. It is also available in the 4- and 5-horsepower sizes (see Fig. 12-17).

Theory of Operation

The capacitor is inserted in series with the start winding (see Fig. 12-18). The phase shift produced by the capacitor is similar to that produced by the capacitor-start, induction-run motor. The run capacitor is usually between 5 and 50 μF. Thus, it is

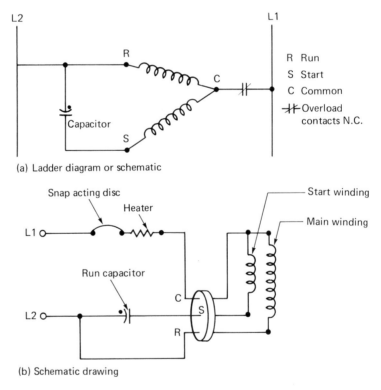

FIGURE 12-17 Permanent split-capacitor motor schematic.

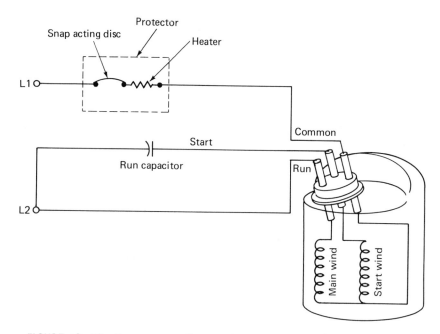

FIGURE 12-18 Permanent split-capacitor compressor schematic. (Courtesy of Tecumseh)

smaller than the start capacitor used in the capacitor-start motor. This also means it will have less starting torque than the capacitor-start motor. However, the torque is enough to start the motor running even with a small load. The capacitor is in the circuit all the time. It is not removed by a relay or any other type of device. It is a permanent part of the circuit.

After the compressor or fan is started and begins to run at speed, the motor produces a counter electromotive force. The cemf builds up to within a few volts of the applied voltage when the motor has reached full speed. As long as the difference between the applied voltage and the cemf is small, very little current flows in the start winding. This is because the capacitor will allow more current to pass as the difference in applied voltage and cemf gets larger, and less current to flow when the voltage difference is small. There is a small difference in voltage at full rpm, so the current in the start winding will be small. The 2 to 4 amperes does not constitute enough to cause damage to the compressor start winding (see Fig. 12-19).

The run capacitor left in the circuit aids in speed regulation. One thing to remember on this type of motor is to make sure the capacitor is placed in the circuit properly. The run capacitor has a *red dot* or some other marking on or near one of the two terminals. This is the outside foil of the capacitor. The run capacitor used here is in a grounded steel case that aids in the dissipation of heat (see Fig. 12-19B).

The outside foil is near the capacitor case. If the insulation breaks down or shorts to the outside foil, it will make contact with the case. Excessive current will

Run capacitor sizes
1/8 hp compressor or motor use a capacitor of 4 or 5 μF
1/6 hp compressor or motor use a capacitor of 4 or 5 μF
1/2 hp compressor or motor use a capacitor of 10 μF
1/2 to 2 hp compressor or motor use a capacitor of 10 to 15 μF
3 hp compressor or motor use 2 capacitors of 10 μF in parallel to equal 20 μF

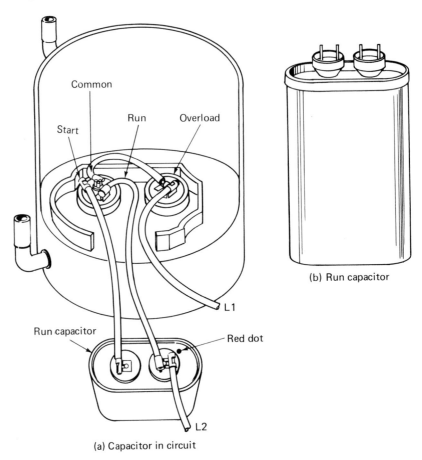

(b) Run capacitor

(a) Capacitor in circuit

FIGURE 12-19 PSC compressor hookup. (Courtesy of Tecumseh)

flow in the circuit when the short occurs. The main objective here is to keep the current as low as possible and have it do the least damage possible. That is why the red dot or mark is always placed where it is supposed to be. In Fig. 12-19, the red dot terminal should be connected to L_2. Then, if it shorts, the foil touches the case, shorts to ground, causes excess current to flow, and trips the circuit breaker or blows the fuse.

However, if the red dot is connected incorrectly, the short will burn out the compressor start winding since it puts the winding directly across the power source by grounding one end of the winding. Make sure the capacitor's red dot is connected to the line side and not to the compressor start winding (see Fig. 12-19A).

Reversibility. The PSC motor can be reversed if it has three wires leading from the case (see Fig. 12-20). To reverse simply connect either side of the capacitor to the line. However, *with compressors it is best to leave them as is.* They have been designed without the possibility of reversing in most cases. To reverse the four-wire-type motor, transpose the black leads as shown in Fig. 12-21.

Uses. Permanent split-capacitor motors may also be used for the fans that are mounted behind the condensers on air-conditioning units. They move the air past the condenser and thus remove the heat from the building being air conditioned. This type of motor can also be varied in speed by changing its windings. A number

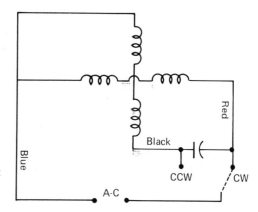

FIGURE 12-20 Permanent split-capacitor three-lead schematic.

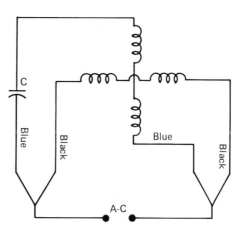

FIGURE 12-21 Permanent split-capacitor four-lead schematic.

of color-coded windings are brought out so they may be connected to a switching arrangement for low, medium, and high speeds.

CAPACITOR-START, CAPACITOR-RUN MOTOR

The capacitor start and run (CSR or CSCR) motor arrangement uses a start capacitor and a run capacitor in parallel with each other and in series with the motor start windings. This motor has high starting torque and runs efficiently. It is used in many refrigeration and air-conditioning applications up to 5 horsepower. A potential relay removes the start capacitor from the circuit after the motor is up to speed. *Potential relays must be accurately matched to the compressor* (see Fig. 12-22). Efficient operation depends on this.

Theory of Operation

The capacitor-start, capacitor-run motor is used on air-conditioning and refrigeration units that need large starting torque. You will find the CSCR motor used on equipment that has an expansion valve and on air-conditioning systems when the permanent-split compressor has trouble starting. *In some cases, the technician adds a capacitor in the field to make the PSC motor start more easily*. This produces the same electrical characteristics as a CSCR type when the additional capacitor is hooked up to the existing arrangement. When the capacitor is added by the technician in the field, it is referred to as a *hard-start kit*.

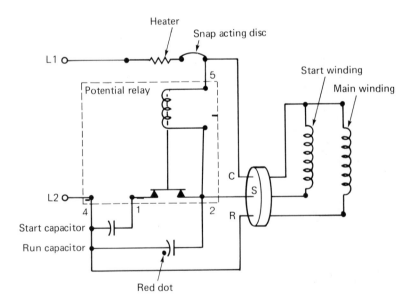

FIGURE 12-22 Capacitor start and run motor schematic. (Courtesy of Tecumseh)

Since the CSCR motor has the additional capacitor added during starting time, some way must be provided to remove it once its purpose has been served. The potential relay is called upon to do the job. This type of relay will be discussed later in this chapter.

The start capacitor is available in sizes up to 600 μF. This is a large capacitor when compared to the run capacitor of up to 75 μF (see Fig. 12-22). The large capacitance value causes a larger phase shift between the start and run winding voltages. As the compressor motor starts to turn, the cemf begins to build. The cemf is present between the S and C terminals of the compressor. The potential relay coil is connected at terminals 2 and 5 of Fig. 12-22. Note the symbol used for the coil of the potential relay. A PR (for potential relay) and a small resistorlike symbol show the coil part of the relay.

When the compressor reaches approximately 75% of its full rpm, the cemf is strong enough to energize the potential relay coil. This pulls the contacts open. Contactors are shown as two parallel lines with a slant line through them with PR underneath. When the potential relay contacts open, the start capacitor is removed from the circuit. This leaves the run capacitor still in the circuit and in series with the start winding. The result is good starting torque and good running efficiency. The efficiency is increased inasmuch as the power factor is brought closer to unity or 1.00.

Reversibility. This type of motor can be reversed by changing the leads from the start winding. It is, however, difficult to do if the compressor is sealed. In open-type motors, it is possible to reverse the direction of rotation by simply reversing the connections of the start windings.

Uses. This type of motor is used on equipment with a need for good starting torque. This includes some types of refrigeration equipment and some hard to start air-conditioning systems.

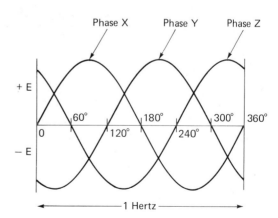

FIGURE 12-23 Three-phase ac waveform.

THREE-PHASE MOTOR

The three-phase motor does not need centrifugal switches or capacitors. The three phases of this type of power generate their own rotating magnetic field when applied to a stator with three sets of windings (see Fig. 12-23). The stator windings are placed 120° apart. The rotor is a form-wound type or a cage type. The squirrel-cage rotor is standard for motors smaller than 1 hp (see Fig. 12-24).

Theory of Operation

For the purpose of identification of phases, note that the three phases are labeled A, B, and C in Fig. 12-25. Phase B is displaced in time from A by 1/3 hertz, and phase C is displaced from phase B by 1/3 hertz. In the stator the different phase windings are placed adjacent to each other so that a B winding is next to an A, a C is next to a B, and then an A is next to a C, and so on around the stator (see Fig. 12-25).

The next step is to picture the magnetic fields produced by the three phases in just one group of A, B, and C windings for 1 hertz. Start with the A phase at its maximum positive peak current value. This A winding is a north pole at its maximum strength. As the cycle progresses, the magnetic pole at A will decrease to zero as the current changes direction. It will become a south pole. The strength of the field will increase until the current reaches its greatest negative value. This produces a maximum-strength south pole, and then a decrease. It then passes through zero or neutral and becomes a maximum north pole at the end of the cycle.

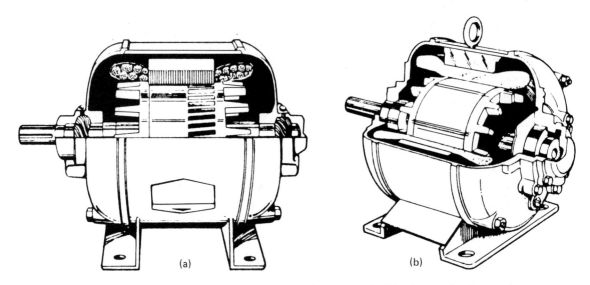

FIGURE 12-24 Cutaway view of a three-phase motor with (A) a half-etched squirrel-cage rotor, and (B) a three-phase motor with a cast rotor.

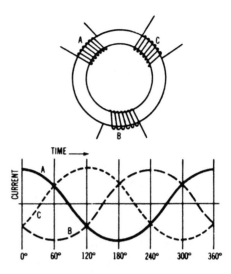

FIGURE 12-25 Three-phase current and coil placement in a motor.

The B phase winding does exactly the same thing, except that the rise and fall of the magnetic fields follow behind the A phase by 1/3 hertz, and the C phase winding magnetic field follows behind B by 1/3 hertz and the A phase by 2/3 hertz.

Assume that you can see only the maximum north poles produced by each phase. Your view of the motor is from the end of a complete stator connected to a three-phase power source. The north poles will move around the stator and appear to be revolving because of the current relationship of the A, B, and C phases.

The three-phase motor has a rotor that consists of steel discs pressed onto the motor shaft. The slots or grooves are filled with aluminum and connected on the ends to form a cage for the electrical current (see Fig. 12-26). The rotating magnetic field in the stator induces current into this electrical cage and thereby sets up north and south poles in the rotor. These north and south poles then follow their opposite members in the stator and the shaft rotates. Polyphase induction motors are often called squirrel-cage motors because of this rotor construction. This rotating field

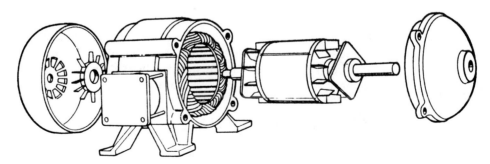

FIGURE 12-26 Three-phase motor with explosion-proof construction.

makes it possible for the motor to start without capacitors or switches. Thus, it is a simpler motor to maintain and operate.

Reversibility. Three-phase motors can be reversed while running. It is very hard on the bearings and the driven machine, but it can be done by reversing any two of the three connections. This is usually done by a switch specially designed for the purpose.

Open Phase. If a three-phase motor develops an open "leg" or one phase (two instead of three wires are coming into the motor terminal with power), it will slow down and hum noticeably. It will, however, continue to run in the same direction. If you try to start it with only two legs (or phases), it will not start but will rotate if started by hand (in fact, it will start in either direction). Once the other phase is connected, it will quickly come up to speed. The loss of one leg is usually due to a fuse in that leg blowing (that is if there are three individual fuses in the three-phase circuit, as is normally the case).

Uses. Three-phase motors are used for machine tools, industrial pumps and fans, air compressors, and air-conditioning equipment. They are recommended wherever a polyphase power supply is available. They provide high starting and breakdown torque with smooth pull-up torque. They are efficient to operate and are designed for 208 to 220/460-volt operation with horsepower ratings from ¼ to the hundreds.

CAPACITOR RATINGS

You have already been introduced to capacitors that are used for starting a motor and for improving the power factor and torque characteristics of the motor. Here we would like to take a look at the ratings and the proper care of capacitors for long-time operation.

Never use a capacitor with a lower rating than specified on the original equipment. The voltage rating and the microfarad rating are important. A higher voltage rating than that specified is always usable. However, a voltage rating lower than that specified can cause damage. Make sure the capacitance marked on the capacitor in microfarads is as specified. Replace with a capacitor of the same size rated in μF, MF, UF, or MFD. All these abbreviations are used to indicate microfarads. See Table 12-1 for a listing of the start capacitors and their ratings for different voltages. Note that the capacitors are not exactly what they are rated for in microfarads. They have a tolerance and the limits are shown.

START CAPACITORS AND BLEEDER RESISTORS

The development of high-power-factor, low-current, single-phase compressor motors that require start and run capacitors used with potential-type relays has created

TABLE 12-1 Ratings and Test Limits for AC Electrolytic Capacitors

Capacity Rating (microfarads)			110-Volt Ratings		125-Volt-Ratings		220-Volt Ratings	
Nominal	Limits	Average	Amps. at Rated Voltage, 60 Hz	Approx. Max. Watts	Amps. at Rated Voltage, 60 Hz	Approx. Max. Watts	Amps. At Rated Voltage, 60 Hz	Approx. Max. Watts
	25– 30	27.5	1.04– 1.24	10.9	1.18– 1.41	14.1	2.07–2.49	43.8
	32– 36	34	1.33– 1.49	13.1	1.51– 1.70	17	2.65–2.99	52.6
	38– 42	40	1.56– 1.74	15.3	1.79– 1.98	19.8	3.15–3.48	61.2
	43– 48	45.5	1.78– 1.99	17.5	2.03– 2.26	22.6	3.57–3.98	70
50	53– 60	56.5	2.20– 2.49	21.9	2.50– 2.83	28.3	4.40–4.98	87.6
60	64– 72	68	2.65– 2.99	26.3	3.02– 3.39	33.9	5.31–5.97	118.2
65	70– 78	74	2.90– 3.23	28.4	3.30– 3.68	36.8	5.81–6.47	128.1
70	75– 84	79.5	3.11– 3.48	30.6	3.53– 3.96	39.6	6.22–6.97	138
80	86– 96	91	3.57– 3.98	35	4.05– 4.52	45.2	7.13–7.96	157.6
90	97–107	102	4.02– 4.44	39.1	4.57– 5.04	50.4	8.05–8.87	175.6
100	108–120	114	4.48– 4.98	43.8	5.09– 5.65	56.5	8.96–9.95	197
115	124–138	131	5.14– 5.72	50.3	5.84– 6.50	65		
135	145–162	154	6.01– 6.72	62.8	6.83– 7.63	85.8		
150	161–180	170	6.68– 7.46	69.8	7.59– 8.48	95.4		
175	189–210	200	7.84– 8.71	81.4	8.91– 9.90	111.4		
180	194–216	205	8.05– 8.96	83.8	9.14–10.18	114.5		
200	216–240	228	8.96– 9.95	93	10.18–11.31	127.2		
215	233–260	247	9.66–10.78	106.7	10.98–12.25	145.5		
225	243–270	257	10.08–11.20	110.9	11.45–12.72	151		
250	270–300	285	11.20–12.44	123.2	12.72–14.14	167.9		
300	324–360	342	13.44–14.93	147.8	15.27–16.96	201.4		
315	340–380	360	14.10–15.76	156				
350	378–420	399	15.68–17.42	172.5				
400	430–480	455	17.83–19.91	197.1				

electrical peculiarities that did not exist in previous designs. In some situations, relay contacts may weld together, causing compressor motor failure. This phenomenon occurs due to the high voltage in the start capacitor discharging (arcing) across the potential relay contacts. To eliminate this, start capacitors are equipped with bleeder resistors across the capacitor terminals (see Fig. 12-27).

Bleeder resistor-equipped capacitors may not be available. Then a 2-watt, 1500-ohm resistor can be soldered across the capacitor terminals. This does not interfere with the operation of the circuit, but does allow the capacitor to discharge across the resistor instead of across the relay switch points. This resistor is called a *bleeder* because it bleeds off the capacitor charge.

Run Capacitors. The marked terminal of run capacitors should be connected to the R terminal of the compressor and thus to L_2. Check the wiring diagram for the correct terminal. The run capacitor is in the circuit whenever the compressor is running. It is an oil-filled electrolytic capacitor that can take continuous use.

If the start capacitor is left in the circuit too long, in some cases the coil of

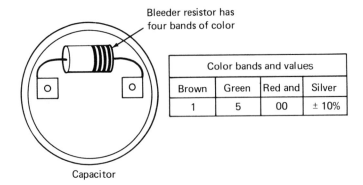

FIGURE 12-27 Bleeder resistor across the capacitor terminals.

the potential relay will open from vibration or use and cause the start capacitor to stay in the circuit longer than 10 to 15 seconds. *This causes the electrolytic capacitor to explode or spew out its contents.*

MOTOR PROTECTORS

In most compressors there is a motor protector (see Fig. 12-28). The overload protector is inserted into the motor windings so that if they overheat the device will

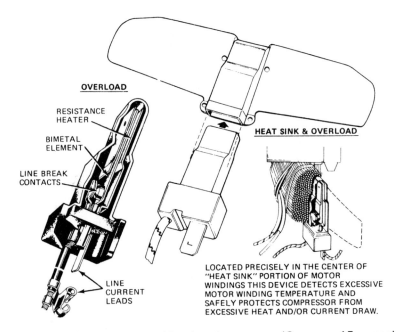

FIGURE 12-28 Motor internal line-break protector. (Courtesy of Tecumseh)

open the contacts of the switch. The bimetal element expands to cause the contacts to remove the power to the windings. These protectors are in addition to any circuit breakers that may be mounted outside the compressor.

COMPRESSOR MOTOR RELAYS

A hermetic compressor motor relay is an automatic switching device designed to disconnect the motor start winding after the motor has come up to running speed (see Fig. 12-29). The two types of motor relays used in refrigeration and air-conditioning compressors are the current type and the potential type.

(a)

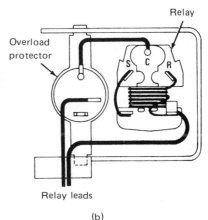

(b)

FIGURE 12-29 A. Location of the overload and relay on a compressor. (Courtesy of Tecumseh) B. Starting relay and overload protector. (Courtesy of Kelvinator)

Current-type Relay

The current-type relay is most often used with small refrigeration compressors up to $\frac{3}{4}$ horsepower. When power is applied to the compressor motor, the relay solenoid coil attracts the relay armature upward. This causes bridging contact and stationary contact to engage (see Fig. 12-30). This energizes the motor start winding. When the compressor motor comes up to running speed, the motor's main winding current is such that the relay solenoid coil de-energizes. This allows the relay contacts to drop open, which disconnects the motor start winding.

One thing to remember about this type of relay is its *mounting*. *It should be mounted in a true vertical position* so that the armature and bridging contact will drop free when the relay solenoid is de-energized.

Potential-type Relay

This relay is generally used with large commercial and air-conditioning compressors (see Fig. 12-31). The motors may be capacitor-start, capacitor-run types up to 5 horsepower. Relay contacts are normally closed. The relay coil is wired across the start winding. It senses voltage change. Start winding voltages increase with motor speed. As the voltage increases to the specific pickup value, the armature pulls up, opening the relay contacts and de-energizing the start windings. After switching, there is still sufficient voltage induced in the start winding to keep the relay coil energized and the relay starting contacts open. When the power is shut off to the motor, the voltage drops to zero. The coil is de-energized and the start contact is reset for the next start.

Many of these relays are *extremely position sensitive*. When changing a compressor relay, care should be taken to install the replacement in the same position

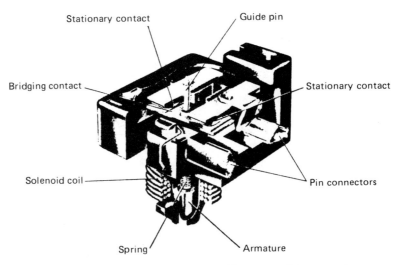

FIGURE 12-30 Current relay. (Courtesy of Tecumseh)

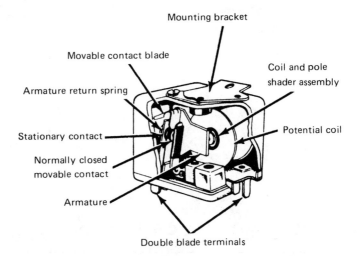

FIGURE 12-31 Potential-type relay. (Courtesy of Tecumseh)

as the original. Never select a replacement relay solely by horsepower or other generalized rating. Select the correct relay from the parts guide book furnished by the manufacturer.

Visual inspection can distinguish the two relays. The current relay has heavy wire for the coil and the potential relay has fine wire for the coil.

REVIEW QUESTIONS

1. On what principle is the induction motor based?
2. What does the skew of a motor mean?
3. What is a shaded-pole motor? Where is it used?
4. What is a split-phase motor? Where is it used?
5. What is a repulsion-start, induction-run motor? Where is it used?
6. What is a capacitor-start motor? Where is it used?
7. What is a permanent split-capacitor (PSC) motor? Where is it used?
8. What is a capacitor-start, capacitor-run motor? Where is it used?
9. What is a three-phase motor? Where is it used?

PERFORMANCE OBJECTIVES

Know how to work around electricity safely.
Know what precautions to take when using portable electrical hand tools.
Know how the ground found circuit interrupter works.
Know where the GFCI is used.
Know the color-coding of wiring systems.

CHAPTER 13

Electrical Safety

Air-conditioning, refrigeration, and heating systems all utilize the convenience of electrical controls. These devices are made with a particular application in mind. Each is designed for a specific purpose. By using low voltage in most instances to control higher voltages, the devices are suited for remote operation since the size of the wire is rather small and inexpensive to install.

Some control devices are designed to protect the technician while repairing or maintaining equipment. Certain safety procedures are necessary to make sure you do not receive a fatal electrical shock. The following should be helpful in making it safe to work on electrically controlled and operated equipment.

SAFETY PRECAUTIONS

It takes very little current to cause physical damage to the human body. In some cases death may result from as little a one-tenth (0.1) of an ampere (see Table 13-1). You may react to a slight shock and then move quickly and come in contact with operating machinery. Involuntary actions caused by electrical shock are more harmful in most instances than the actual mild shock. It is very easy to become careless. Just keep in mind that the human body has a skin resistance of between 400,000 and 800,000 ohms. This can be used in an Ohm's law formula to determine the current

$$I = E/R$$

$$I = 120/400,000 = 0.3 \text{ milliamperes}$$

TABLE 13-1
Physiological Effects of Electric Currents[a]

	Readings (mA)	Effects
Safe current values	1 or less	Causes no sensation—not felt.
	1 to 8	Sensation of shock, not painful; individual can let go at will since muscular control is not lost.
Unsafe current values	8 to 15	Painful shock; individual can let go at will since muscular control is not lost.
	15 to 20	Painful shock; control of adjacent muscles lost; victim cannot let go.
	20 to 50	Painful, severe muscular contractions; breathing difficult.
	50 to 100	Ventricular fibrillation, a heart condition that can result in instant death is *possible*.
	100 to 200	Ventricular fibrillation occurs.
	200 and over	Severe burns, severe muscular contractions, so severe that chest muscles clamp the heart and stop it for the duration of the shock. (This prevents ventricular fibrillation.)

[a]Information provided by National Safety Council.

This is not enough to cause you to feel it. However, if you have wet hands or make contact with part of the body that is not dry, you may have a body resistance as low as 50,000 ohms and receive a current of 2.4 milliamperes. That is enough to cause you to jump back from a slight tingle. The backward movement on a roof may be enough to cause you to fall a great distance. You may also drop the equipment or tools you are working with and cause further damage. If you are working around 240 volts, the danger is greater. At 240 volts a contact resistance of 50,000 ohms produces 4.8 milliamperes or enough to cause a heavy shock with tightening of the muscles. A low body resistance of say 200 ohms can be fatal. This can happen whenever the skin is wet or you are standing on a wet surface. With 200 ohms resistance and 120 volts, you will receive 600 milliamperes, enough to cause death.

MAIN SWITCHES

If you are working on equipment with the power on, be sure you are standing on a dry surface and your hands do not make contact with anything other than the probes being used to make measurements. It is always good practice when working around live circuits to *keep one hand in your pocket* so that you will not complete the path across your chest. It is, of course, always better to have the power off when working on equipment. A good procedure is to turn *off* the main circuit breakers and put a sign on them so that no one turns them on without your knowledge. If there is a control box or distribution panel with a lock on it, make sure the lock is in place and locked with the lever in the down or off position (see Figs. 13-1 and 13-6). Fusing of the circuits is shown in Fig. 13-1.

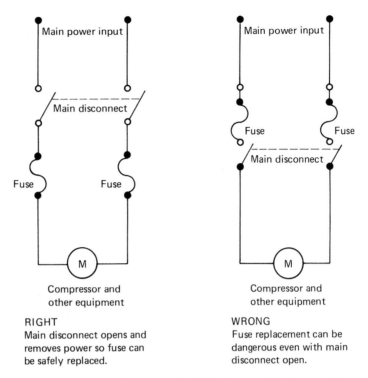

FIGURE 13-1 Main disconnect should be located before working on equipment.

PORTABLE ELECTRICAL TOOLS

Grounding portable tools aids in preventing shocks and damage to equipment under certain conditions. A grounding conductor has a white-colored jacket in a two- or three-wire cable (neutral wire). It is terminated to the white- or silver-colored terminal in a plug cap or connector. And it is terminated at the neutral bar in the distribution box. Keep in mind that gray is also used to color code wires in industrial and commercial installations. Gray wires are also used for grounding.

When there is an electrical fault that allows the hot line to contact the metal housing of electrical equipment, in a typical two-wire system, or some other ungrounded conductors, any person who touches that equipment or conductor will receive a shock. The person completes the circuit from the hot line to the ground and current passes through the body. Because a body is not a good conductor, the current is not high enough to blow the fuse. Thus, the current continues to pass through the body as long as the body remains in contact with the equipment (see Fig. 13-2).

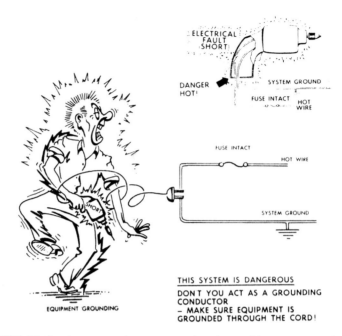

FIGURE 13-2 A person who contacts the charged housing of a drill or piece of equipment becomes the conductor in a short circuit to ground. (Courtesy of the National Safety Council)

A grounding conductor, or equipment ground, is a wire attached to the housing or other conductive parts of electrical equipment that are not normally energized to carry current from them to the ground. Thus, if a person touches a part that is accidentally energized, there will be no shock, because the grounding line furnishes a much lower resistance path to the ground (see Fig. 13-3). The high current passes through the wire conductor and blows the fuses and stops the current. In normal operation, a grounding conductor does not carry current.

The grounding conductor in a three-wire conductor cable has a green jacket. The grounding conductor is always terminated at the green-colored hexhead screw on the cap or connector. It utilizes either a green-colored conductor or a metallic conductor as its path to ground. In Canada, this conductor is referred to as the *earthing conductor,* which is somewhat more descriptive and helpful in distinguishing between grounding conductors and neutral wires, or grounded conductors.

Ground-fault Circuit Interrupters

A safety device that should be used by all technicians in the field who do not have time to test out each circuit before it is used is the ground-fault circuit interrupter (GFCI). It is designed to protect you from shock.

PORTABLE ELECTRICAL TOOLS   251

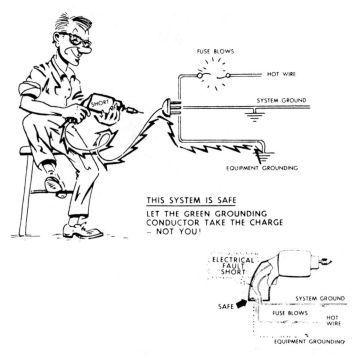

FIGURE 13-3 Properly wired circuit causes the shorted equipment to be shorted to ground instead of through the person to ground. (Courtesy of the National Safety Council)

The differential ground-fault interrupter, available in various modifications, has current-carrying conductors passing through the circular iron core of a doughnut-shaped differential transformer. As long as all the electricity passes through the transformer, the differential transformer is not affected and will not trigger the sensing circuit. If a portion of the current flows to ground and through the fault-detector line, however, the flow of electricity through the sensing windings of the differential transformer causes the sensing circuit to open the circuit breaker. These devices can be arranged to interrupt a circuit for currents of as little as 5 milliamperes flowing to ground (see Fig. 13-4).

Another design is the *isolation-type* ground-fault interrupter. This unit combines the safety of an isolation system with the response of an electronic sensing circuit. In this setup, an isolating transformer provides an inductive coupling between load and line. Both the hot and neutral wires are connected to the isolating transformer. There is no continuous wire connected between.

In the latter type of interrupter, a ground fault must pass through the electronic sensing circuit, which has sufficient resistance to limit current flow to as low as 2 milliamperes, well below the level of human perception.

FIGURE 13-4 Ground-fault circuit interrupter (GFCI) used for portable tools.

TYPES OF CIRCUIT PROTECTORS

There are two types of circuit protectors made in the circuit breaker configuration. They are circuit breakers that work on the heating principle. A bimetallic strip is heated by having circuit current pass through it. When too much current flows, the strip is overheated. The expansion of the strip causes the breaker to trip, thereby causing the circuit to be opened (see Fig. 13-5). This type is slower to cool down than the magnetic type. It takes a little longer for it to cool to a point where it can be reset. If the overload still exists, the breaker will trip again.

The *magnetic type* uses a coil to operate. The circuit current is drawn through the coil. If too much current flows, the magnetic properties of the coil cause the circuit breaker to trip and open the circuit. This type is quicker in terms of being able to reset it. The overload must, of course, be removed before it will remain in the reset position.

In most cases where the power is 240 volts single-phase, the circuit breakers will be locked together with a pin. If one side of the power circuit is opened, the other circuit is also tripped since they are tied together physically.

In three-phase circuits, the three circuit breakers are tied together so that when one trips all three circuits are disconnected from the power line.

If air-conditioning or refrigeration equipment is served by a separate box, make sure the lock is in place and the handle is in the down or *off* position before making an inspection or working on the equipment (see Fig. 13-6).

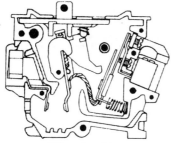

Tripped position
- Contacts open — no current flow
- Handle stationary when tripped

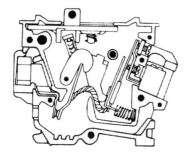

"On" position
- Contacts closed — current on
- Handle in "On" position (shows "On").

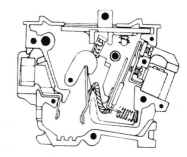

"Off" position
- Contacts open — no current flow
- Handle in "Off" position (shows "Off")

To restore service when fault is cleared you simply move operating handle to "Off" position and then to "On".

FIGURE 13-5 Cutaway view of a circuit breaker in various positions. (Courtesy of Wadsworth)

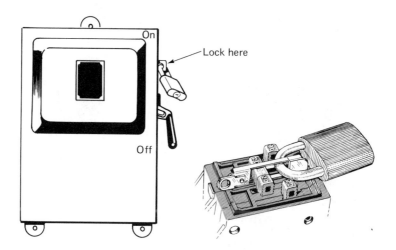

FIGURE 13-6 Lock the box by using a lock and keep the key until you need to unlock it.

REVIEW QUESTIONS

1. How much current does it take to kill?
2. What precautions should you take when using portable electric hand tools?
3. What is a ground-fault circuit interrupter?
4. Where is the GFCI used?
5. What is the color of the ground wire in a two-wire system?
6. What is the other color used for ground in an industrial or commercial installation?
7. Why should you lock the off switch when working with air-conditioning, refrigeration, or heating equipment?

PERFORMANCE OBJECTIVES

Know how power relays work.
Know how time delay relays work.
Know the most common types of thermostats.
Know the advantages of microprocessor thermostats.
Know how a cold anticipator works and how to adjust it.
Know how a heat anticipator works and how to adjust it.
Know how to adjust a limit switch.
Know how water tower controls operate.
Know how pressure control switches operate.

CHAPTER 14

Control Devices

Many types of controls are available for use on air-conditioning, refrigeration, and heating equipment. They come in many sizes and shapes and do the job well for a period of time, but they all require periodic inspections, repairs, and replacement. One of these controls is the power relay. It is one of the simplest controls used to control compressors for refrigeration and air-conditioning purposes.

POWER RELAY

The power relay is also referred to as the *main conductor*. It is used to apply the main line voltage to the motor circuit. The coil of the relay is usually operated by voltages lower than the line provides. This means that it uses a transformer for the lower control voltages (see Fig. 14-1). The symbols used for this type of relay are shown in Fig. 14-1.

Magnetic contactors are normally used for starting polyphase motors, either squirrel cage or single phase. Contactors may be connected at any convenient point in the main circuit between the fuses and the motor. Small control wires (using low voltage) may be run between the contactor and the point of control.

Motor Start Relay

Relays are a necessary part of many control and pilot-light circuits. They are similar in design to contactors, but are generally lighter in construction so they carry smaller currents.

Compressors used for household refrigerators, freezers, dehumidifiers, vend-

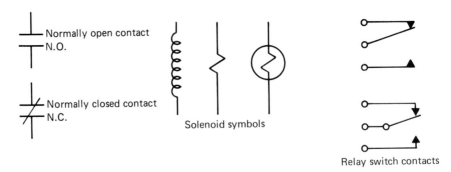

FIGURE 14-1 Symbols for the main contactor or power relay.

ing machines, and water coolers have the capacitor-start, induction-run type of motor. This type of compressor may have a circuit that resembles Fig. 14-2. When the compressor is turned on by the thermostat demanding action, the relay is closed and the start winding is in the circuit. Once the motor comes up to about 75% of rated speed, there is enough current flow through the relay coil to cause it to energize, and it pulls the contacts of the relay open, thereby taking the start capacitor and start winding out of the circuit. This allows the motor to run with one winding as designed.

Figure 14-3 shows the current type of relay. This is generally used with small refrigeration compressors up to 3/4 horsepower. Figure 14-4 shows the potential type of relay. This is generally used with large commercial and air-conditioning compressors up to 5 horsepower.

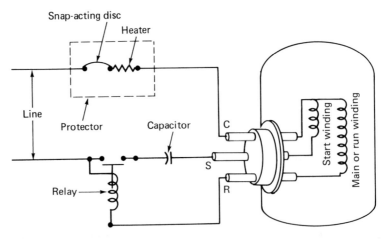

FIGURE 14-2 Capacitor-start, induction-run motor for a compressor with the potential relay used to take out the start winding once the motor comes up to speed.

POWER RELAY 259

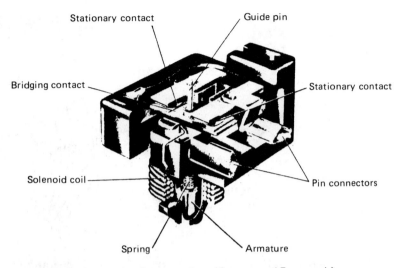

FIGURE 14-3 Current relay. (Courtesy of Tecumseh)

Protection of the motor against prolonged overload is accomplished by time-limit overload relays. They are operative during the starting period and running period. Relay action is delayed long enough to take care of the heavy starting currents and momentary overloads without tripping.

Relays with More than One Contact

Some power relays are made with more than one set of contacts. They are used to cause a sequence of events to take place. The contacts can be wired into a circuit

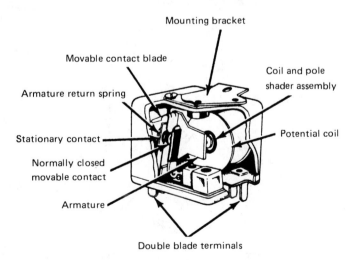

FIGURE 14-4 Potential relay. (Courtesy of Tecumseh)

260 CONTROL DEVICES

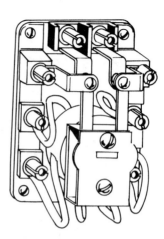

FIGURE 14-5 Relay with more than one set of contacts.

that controls functions other than the on-off operation of the compressor motor (see Fig. 14-5).

Thermal Overload Protectors

Motors for commercial units are normally protected by a bimetallic switch. The switch is operated on the heat principle. This is a built-in motor overload protector (see Fig. 14-6). It limits the motor winding temperature to a safe value. In its sim-

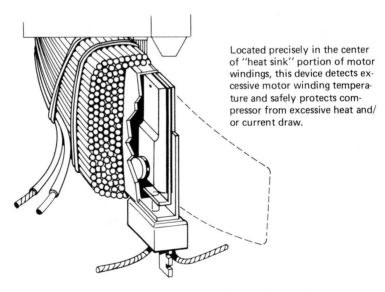

Located precisely in the center of "heat sink" portion of motor windings, this device detects excessive motor winding temperature and safely protects compressor from excessive heat and/or current draw.

FIGURE 14-6 Motor protector inserted in the windings of the compressor. (Courtesy of Tecumseh)

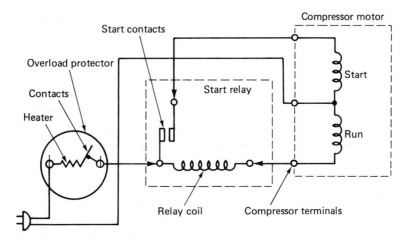

FIGURE 14-7 Domestic refrigerator circuit showing the start contacts and relay coil, as well as the overload protector.

plest form, the switch or motor protector consists essentially of a bimetal switch mechanism that is permanently mounted and connected in series with the motor circuit (see Fig. 14-7). Figure 14-8 shows how the external line-break overload operates.

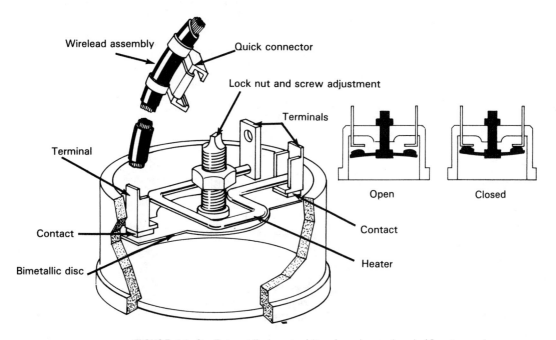

FIGURE 14-8 Externally located line-break overload. (Courtesy of Tecumseh)

262 CONTROL DEVICES

TIME-DELAY RELAYS

In time-delay relays, bimetallic strips are heated with an electrical resistance mounted near or around them. The strips expand when heated. When they expand, they make contact and complete the circuit with their contacts closed (see Fig. 14-9). The time delay can be adjusted by the resistance of the heater unit. This type of unit is different from that shown as a protector in Fig. 14-7. The heating element in Fig. 14-7 causes the circuit to open and protect the motor. The time-delay relay is used to make sure that certain things take place within the refrigeration cycle before another is commenced.

SOLENOIDS

Solenoid valves are used in many heating and cooling applications. They are electrically operated. A solenoid valve, when connected as in Fig 14-10, remains open when current is supplied to it. It closes when the current is turned off. In general,

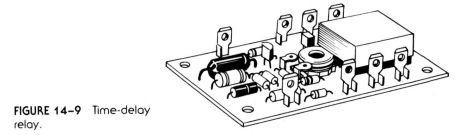

FIGURE 14-9 Time-delay relay.

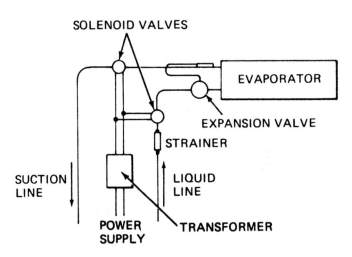

FIGURE 14-10 Solenoid valves connected in the suction and liquid evaporator lines of a refrigeration system.

solenoid valves are used to control the liquid refrigerant flow into the expansion valve or the refrigerant gas flow from the evaporator when it or the fixture it is controlling reaches the desired temperature. The most common application of the solenoid valve is in the liquid line, and it operates with a thermostat (see Fig. 14-11).

The solenoid shown in Fig. 14-12 controls the flow of natural gas in a hot-air furnace. Note how the coil is wound around the plunger. The plunger is the core of the solenoid. It has a tendency to be sucked into the coil whenever the coil is energized by current flowing through it. The electromagnetic effect causes the plunger to be attracted upward into the coil area. When the plunger is moved upward by the pull of the electromagnet, the soft disc (10) is pulled upward, allowing gas to flow through the valve. This basic technique is used to control water, gasoline, oil, or any other liquid or gas.

THERMOSTATS

Temperature control by using thermostats is common to both heating and cooling equipment. Thermostats are used to control heating circuits that cause furnaces and boilers to operate and provide heat. Thermostats are also used to control cooling equipment and refrigeration units. Each of these purposes may have its own specially designed thermostat or may use the same one. For instance, in the home you use the same thermostat to control the furnace and the air-conditioning unit.

Bellows-type Thermostat

On modern condensing units, low-pressure control switches are largely superseded by thermostatic control switches. A thermostatic control consists of three main parts: a bulb, a capillary tube, and a power element or switch. The bulb is attached to the evaporator in a manner that assures contact with the evaporator. It may contain a volatile liquid, such as a refrigerant. The bulb is connected to the power element by means of a small capillary tube (see Fig. 14-13).

Operation of the bellows is provided by a change in temperature. Or the opera-

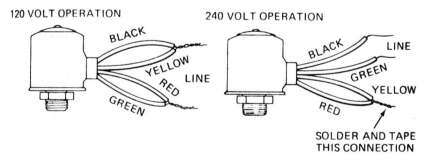

FIGURE 14-11 Solenoid valves. Note color-coded wires. (Courtesy of General Controls)

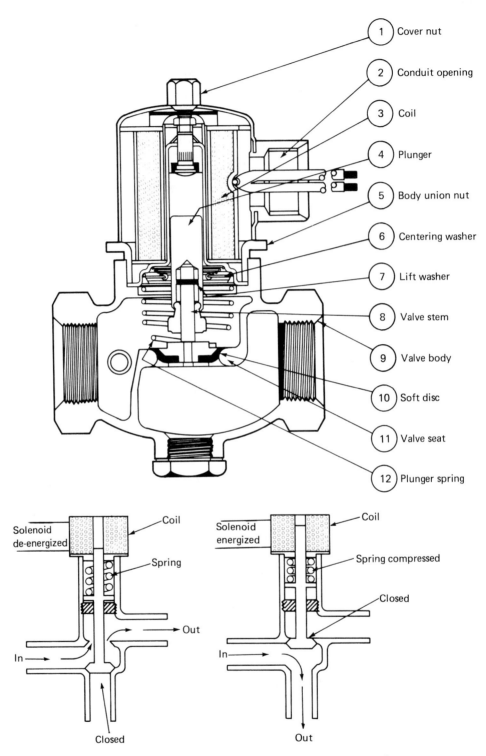

FIGURE 14-12 Solenoid used for controlling natural gas flow to a furnace. (Courtesy of Honeywell)

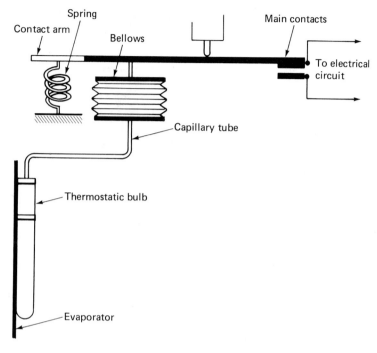

FIGURE 14-13 Bellows-type switch.

tion of the thermostatic control switch is such that, as the evaporator temperature increases, the bulb temperature also increases. This raises the pressure of the thermostatic liquid vapor. This, in turn, causes the bellows to expand and actuate an electrical contact. The contact closes the motor circuit, and the motor and compressor start operating. As the evaporator temperature decreases, the bulb becomes colder and the pressure decreases to the point where the bellows contracts sufficiently to open the electrical contacts, thus turning off the motor circuits. In this manner, the condensing unit is entirely automatic. Thus, it is able to produce exactly the amount of refrigeration needed to meet any normal operating condition.

Bimetallic-type Thermostat

Temperature changes can cause a bimetallic strip to expand or contract in step with changes in temperature. These thermostats are designed for the control of heating and cooling in air-conditioning units, refrigeration storage rooms, greenhouses, fan coils, blast coils, and similar units. This is the type used in most homes for control of the central air-conditioning and central heating system.

Figure 14-14 shows how the bimetallic strip thermostat works. Two metals, each having a different coefficient of expansion, are welded together to form a bimetallic unit or blade. With the blade securely anchored at one end, a circuit is

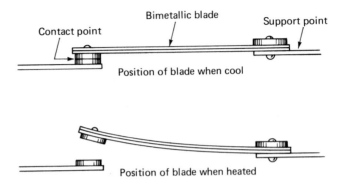

FIGURE 14-14 Bimetallic strip used in a thermostat.

formed and the contact points are closed. This allows the passage of an electric current through the closed points. Because an electric current provides heat in its passage through the bimetallic blade, the metals in the blade begin to expand. However, they expand at a different rate. The metals in the blade are so arranged that the one with a greater coefficient of expansion is placed at the bottom of the unit. After a certain time, the operating temperature is reached and the contact points become separated. This disconnects the device from its power source. After a short period, the contact blade will again become sufficiently cooled to cause the contact point to join, thus reestablishing the circuit and permitting the current again to actuate the circuit. The cycle is repeated over and over again. In this way, the bimetallic thermostat prevents the temperature from rising too high or dropping too low.

Heating and Cooling Thermostats

Some thermostats can be used for both heating and cooling. The thermostat shown in Fig. 14-15 is such a device. The basic thermostat element has a permanently sealed, magnetic SPDT switch. The thermostat element plugs into the subbase and contains the heat anticipation, the magnetic switching, and a room temperature thermometer. The subbase unit contains fixed cool anticipation and circuitry. This thermostat is used with 24 volts ac. In this case, the thermostatic element (bimetal) does not make direct contact with the electrical circuit. Instead, the expansion of the bimetal causes a magnet to move. This, in turn, causes the switch to close or open. Figure 14-16 shows that the bimetal is not in the electrical circuit.

Mercury Contacts. Some thermostats use the expanding bimetal arrangement to cause a tube of mercury to move. As the mercury moves in the tube, it comes in contact with two wires inserted into the glass tube. The mercury contacts the two wires and causes a current to flow between the wires. The mercury completes the electrical circuit. This type of thermostat needs to be so arranged that the tube of

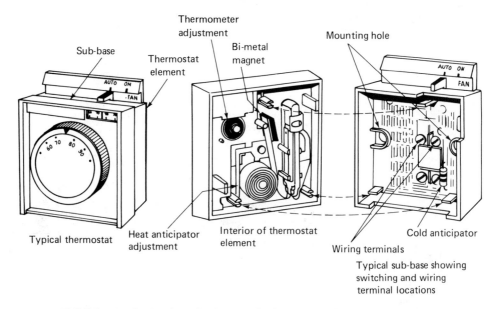

FIGURE 14-15 Modern thermostat for heating and cooling of a house. (Courtesy of General Controls)

mercury is pivoted and can be moved by the expanding or contracting bimetal strip, which exerts or releases pressure on the tube of mercury.

Thermostats used in home air-conditioning and heating systems are now equipped with mercury contacts (see Fig. 14-17). They are made so that the mercury contacts two wires that control the air conditioning in one position and two wires that control the heating system in the other position (see Fig. 14-18).

The advantage of the mercury bulb type of switch is the elimination of switch contact points. Contact points are in need of constant attention. In most cases the dust from the air will eventually cause them to function improperly. It is necessary to clean the points by running a piece of clean paper through them to remove the dust particles and arcing residue. Since the mercury type is sealed and the arcing created on make and break of the circuit simply causes the mercury to vaporize slightly and then return to a liquid state, it provides a trouble-free switching operation.

MICROPROCESSOR THERMOSTATS

Semiconductor technology has produced another means of more accurately controlling air-conditioning and heating systems to provide better regulated temperatures in the home, office, and business. The microprocessor makes use of the semiconductor chip or integrated circuit discussed earlier in Chapter 11. All the external

FIGURE 14-16 Wiring diagram for a thermostat.

connections are the same as for any other type of thermostat. Only the internal circuitry has changed to provide a better regulated temperature and a variety of operations that allow you to set it for any energy-saving program desired (see Fig. 14-19). Unless a battery is included, it does not retain the program in most instances, and the clock, if there is one on the unit, has to be reset each time the power goes off.

MICROPROCESSOR THERMOSTATS 269

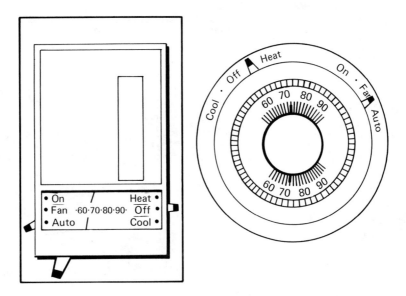

FIGURE 14-17 Thermostat for air conditioning and heating.

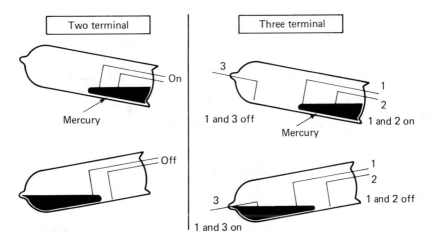

FIGURE 14-18 Mercury-switch operation.

270 CONTROL DEVICES

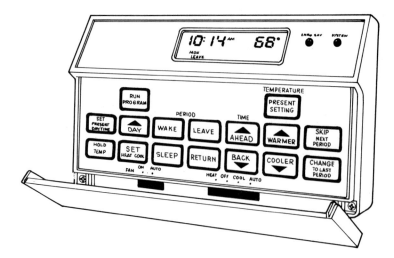

FIGURE 14-19 Microprocessor used for air-conditioning and furnace control in a home.

THERMOSTAT ADJUSTMENTS

In Fig. 14-16, a cold anticipation and a heat anticipation adjustment are placed in the thermostat circuit. The heat anticipation control is placed in series with the switch. The cold anticipation resistor is placed in shunt or parallel with the switch. Thus, when the switch is closed the shunt is shorted out.

Heat Anticipator

The reason for the heat anticipator is to limit the degree of swing between turning on the furnace and the temperature of the room. It is a resistance heater element that is inserted in series with the thermostat line that runs to the heat contactor coil. When the thermostat contacts are closed, current flows through the resistor. This causes it to heat up. The heat generated by the resistor causes the thermostat to open slightly before the desired room temperature is reached by the heating system. This allows the heat in the plenum of the furnace to continue to heat the room. Thus, the resistor aids the thermostat in anticipating the amount of heat that will be provided to the room by using the heat already produced in the plenum.

Cold Anticipator

The cold anticipator is a fixed resistor and is not adjustable. It heats the bimetallic coil that operates the points whenever the air-conditioner compressor is not on. When the compressor is on, the resistor is shorted out by the thermostat points being closed. The heating of the coil while the points are open causes it to close a little earlier than if it waited for the room to heat up sufficiently to cause it to turn on. This way the heat produced by the anticipator resistor causes the compressor to turn on a little before the thermostat would have normally told it to do so. By turning it on before the room has reached the selected temperature, the anticipator causes the temperature swing in the room to be reduced and makes it more comfortable.

LIMIT SWITCHES

Many types of switches are used to limit the amount of heat produced in a furnace. The upper limit has to be controlled so that the furnace does not cause fires by overheating. Limit switches take various forms depending on the manufacturer. However, Fig. 14-20 shows a typical switch and how it works. This is a combination of fan and limit controller that combines the functions of a fan controller and a limit controller in a single unit. One sensing element is used for both controls.

Combination controllers are wired in much the same way as individual controls. These combined controls can be used on line voltage, low voltage, or self-energizing millivolt systems.

Figure 14-21 shows the fluid-filled type of capillary tube used in a limit switch. The one shown in Fig. 14-20 is the bimetal type that twists as it heats up, causing the control unit to move. These limit switches are placed in the plenum of the furnace to control when the fan goes on and off; when the plenum has reached the desired

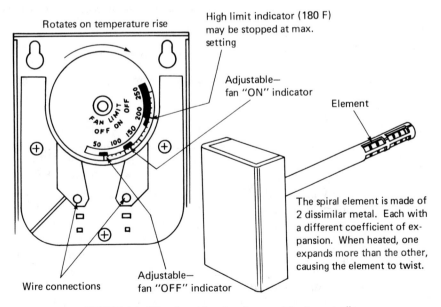

FIGURE 14-20 Combination fan and limit controller.

272 CONTROL DEVICES

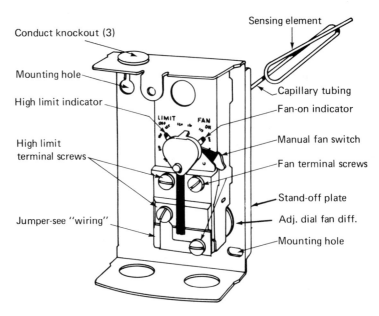

FIGURE 14-21 Combination fan and limit controller.

temperature, it turns off the solenoid and shuts off the flow of natural gas to the burner. Limit switches of a slightly different configuration are also used for electrical strip heaters. They may also be of the low voltage (24 volts) or line voltage type.

PRESSURE CONTROL SWITCHES

Another safety feature for air-conditioning units with a compressor and condenser are pressure-controlled switches. These switches are wired into the circuit to protect the system in case the system develops a leak. If a leak develops, it is possible to draw in moisture and air and damage the whole system. If the pressure builds too high, it can cause a rupture of any of the joints or weaker points in the system.

A low-voltage (24-volt) relay is wired into the 240-volt line that supplies the compressor motor. The relay contacts are wired into the supply line for the motor (see Fig. 14-22). The solenoid of the relay is wired in series with two pressure-operated switches. If the pressure builds too high, the high switch will open and cause the solenoid to de-energize. If this happens, it causes the contacts of the relay to open. This removes power from the compressor motor. If the low-pressure switch opens, it will do the same thing. This way the compressor is protected from both high and low pressure causing damage to the system.

Both manual and automatic controls are available. Automatic controls reset when the pressure stabilizes in the system. If it is not stabilized, it will again turn the system off and keep recycling until it reaches the design pressure.

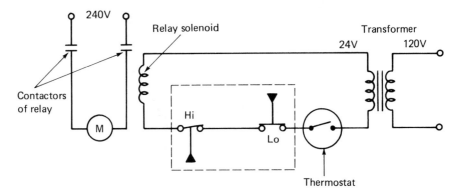

FIGURE 14-22 Pressure-operated switches control the compressor.

WATER TOWER CONTROLS

Temperature controls for refrigerating service are designed to maintain adequate head pressure with evaporative condensers and cooling towers. Low refrigerant head pressure, caused by abnormally low cooling water temperature, reduces the capacity of the refrigeration system.

Two systems of control for mechanical and atmospheric draft towers and evaporative condensers are shown in Figs. 14-23 and 14-24. The control opens the

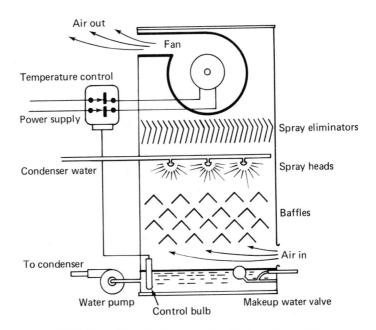

FIGURE 14-23 Cooling tower with forced-air draft.

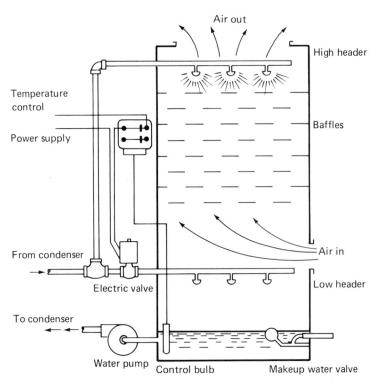

FIGURE 14-24 Cooling tower with atmospheric draft cooling.

contacts when the temperature drops. These contacts are wired in series with the fan motor. Or they can be wired to the pilot of a fan-motor controller. Opening the contacts stops the fan when the cooling water temperature falls to a predetermined minimum value. This value corresponds to the minimum head pressure for proper operation. In the control system shown in Fig. 14-24, the contacts close on a temperature drop and are wired in series with a normally closed motorized valve or a solenoid valve. The contacts open the valve when low cooling temperature occurs. The cooling water then flows through a low header in the atmospheric tower. This reduces its cooling effect and the head pressure increases.

Float switches are used to control the level of water in the cooling tower. Automatic float switches provide automatic control for motors operating tank or sump pumps. They are built in several styles and can be supplied with several types of accessories that provide rod or chain operation and either wall or floor mounting.

A sensor system may also be used. There are hundreds of sensor types. They usually sense the level of water by using two probes. When the water contacts the probes, it causes a small electrical current (at low voltages) to flow and energize a solenoid or relay that in turn causes the water to be turned off. When the level of water is below the two probes and a complete circuit is not available, the normally closed relay contacts are closed by de-energizing the relay. This causes the water solenoid to be energized. This allows makeup water to flow into the cooling tower until it reaches the point where the probes are immersed in water and the cycle is repeated.

REVIEW QUESTIONS

1. What is another name for a power relay?
2. What is the function of the power relay?
3. What is the time-delay relay used for?
4. What is the most common type of thermostat used on air conditioners?
5. Why are mercury contacts better?
6. What is the advantage of a microprocessor thermostat?
7. What is a heat anticipator?
8. Why is a heat anticipator needed?
9. How does the heat anticipator work?
10. What is a limit switch?
11. How does the limit switch work?
12. Where would you find a limit switch?

PERFORMANCE OBJECTIVES

Know the parts of a basic electrical heating system.
Know how to read a ladder diagram.
Know the difference between field wiring and factory wiring.
Know how to trace the low voltage system in a heating unit.
Know how to identify parts on a printed circuit board for a furnace.
Know how a heat pump works.
Know how a refrigerator can be used to generate heat.
Know why supplemental heaters are sometimes needed.
Know the gas/air mixture is ignited in a Pulse™ furnace.
Understand why the Pulse furnace is so efficient.
Know how to troubleshoot the Pulse furnace using the flow chart provided by the manufacturer.

CHAPTER 15

Heating Circuits

Hot-air furnaces are self-contained and self-enclosed units. They are usually centrally located within a building or house. Their purpose is to make sure the temperature of the interior of the structure is maintained at a comfortable level throughout. The design of the furnace is determined by the type of fuel used to fire it. Cool air enters the furnace and is heated as it comes in contact with the hot, metal heating surfaces. As the air becomes warmer, it also becomes lighter, which causes it to rise. The warmer, lighter air continues to rise until it is either discharged directly into a room, as in the pipeless gravity system, or is carried through a duct system to warm-air outlets located at some distance from the furnace.

After the hot air loses its heat, it becomes cooler and heavier. Its increased weight causes it to fall back to the furnace, where it is reheated and repeats the cycle. This is a very simplified description of the operating principles involved in hot-air heating. And it is especially typical of those involved in gravity heating systems. The forced-air system relies on a blower to make sure the air is delivered to its intended location. The blower also causes the return air to move back to the furnace faster than with the gravity system.

With the addition of a blower to the system, there must be some way of turning the blower on when needed to move the air and to turn it off when the room has reached the desired temperature. Thus, electrical controls are needed to control the blower action.

BASIC GAS FURNACE OPERATION

The gas furnace is the simplest to operate and understand. Therefore, we will use it here to look at a typical heating system. This type of natural-gas furnace is used to heat millions of homes in the United States.

Figure 15-1 is a simple circuit needed to control the furnace with a blower. Note the location of the blower switch and the limit switch. The transformer provides low voltage for control of the gas solenoid. If the limit switch opens (it is shown in a closed position), there is no power to the transformer and the gas solenoid cannot energize. This is a safety precaution because the limit switch will open if the furnace gets too hot. When the thermostat closes, it provides 24 volts to the gas solenoid, which energizes and turns on the gas. The gas is ignited by the pilot light and provides heat to the plenum of the furnace. When the air in the plenum reaches 120°F, the fan switch closes and the fan starts. The fan switch provides the necessary 120 volts to the fan motor for it to operate.

Once the room has heated up to the desired thermostat setting, the thermostat opens. When it opens, the gas solenoid is de-energized, and the spring action of the solenoid causes it to close off the gas supply, thereby turning off the source of heat. When the plenum on top of the furnace reaches 90°F, the blower switch opens and turns off the blower. As the room cools down, causing the thermostat to once again close, the cycle starts over again. The gas solenoid opens to let in the gas and the pilot light ignites it. The heat causes the temperature to rise in the plenum above the limit switch's setting and the switch closes to start the blower. Once the thermostat has been satisfied, it opens and causes the gas solenoid to turn off the gas

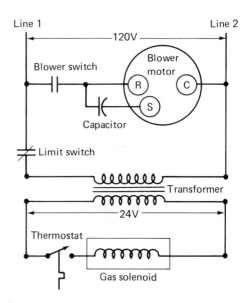

FIGURE 15-1 Simple one-stage furnace-control system.

supply. The blower continues to run until the temperature in the plenum reaches 90°F and it turns off the blower by opening. This cycle is repeated over and over again to keep the room or house at a desired temperature.

BASIC ELECTRIC HEATING SYSTEM

Electric-fired heat is the only heat produced almost as fast as the thermostat calls for it. It is almost instantaneous. There are no heat exchangers to warm up. The heating elements start producing heat the moment the thermostat calls for it. A number of types of electric-fired furnaces are available. They can be bought in 5- to 35-kilowatt sizes. The outside looks almost the same as the gas-fired furnace. The heating elements are located where the heat exchangers would normally be located. Since they draw high amperage, they need electrical controls that can take the high currents.

The operating principle is simple. The temperature selector on the thermostat is set for the desired temperature. When the temperature in the room falls below this setting, the thermostat calls for heat and causes the first heating circuit in the furnace to be turned on. There is generally a delay of about 15 seconds before the furnace blower starts. This prevents the blower from circulating cool air in the winter. After about 30 seconds, the second heating circuit is turned on. The other circuits are turned on one by one in a timed sequence.

When the temperature reaches the desired level, the thermostat opens. After a short time, the first heating circuit is shut off. The others are shut off one by one in a timed sequence. The blower continues to operate until the air temperature in the furnace drops below a specified temperature.

Basic Operation

In Figure 15-2 the electrical heating system has a few more controls than the basic gas-fired furnace. The low-resistance element used for heating draws a lot of current, so the main contacts have to be of sufficient size to handle the current.

The thermostat closes and completes the circuit to the heating sequencer coil. The sequencer coil heats the bimetal strip that causes the main contacts to close. Once the main contacts are closed, the heating element is in the circuit and across the 240-volt line. The auxiliary contacts will also close at the same time as the main contacts. When the auxiliary contacts close, they complete the low-voltage circuit to the fan relay. The furnace fan will be turned on at this time.

Once the thermostat has been satisfied, it opens. This allows the heating sequencer coil to cool down slowly. Thus, the main contacts do not open immediately to remove the heating element from the line. So the furnace continues to produce heat after the thermostat has been satisfied. The bimetal cools down in about 2 minutes. Once it cools, it opens the main and auxiliary contacts, which removes the heating element from the line and also stops the fan motor. After the room cools

HEATING CIRCUITS

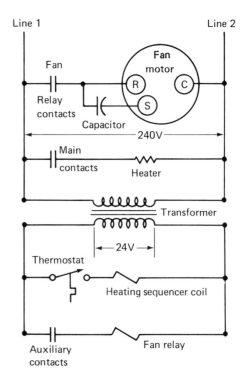

FIGURE 15-2 Ladder diagram for a hot-air furnace.

down below the thermostat setting, the thermostat closes and starts the sequence all over again.

LADDER DIAGRAMS

Electrical schematics are used to make it simple to trace the circuits of various devices. Some of these can appear complicated, but they are usually very simple when you start at the beginning and wind up at the end. The beginning is one side of the power line and the end is the other side of the line. What happens in between is that a number of switches are used to make sure the device turns on or off when it is supposed to cool, freeze, or heat.

The ladder diagram makes it easier to see how these devices are wired. It consists of two wires drawn parallel and representing the main power source. Along each side you find connections. By simply looking from left to right, you are able to trace the required power for the device. Symbols are used to represent the devices. There is usually a legend on the side of the diagram to tell you, for example, that CC means compressor contactor, EFR means evaporator fan relay, and HR means heating relay (see Fig. 15-3).

Take a look at the thermostat in Fig. 15-3. The location of the switch determines whether the evaporator fan relay coil is energized, the compressor contactor

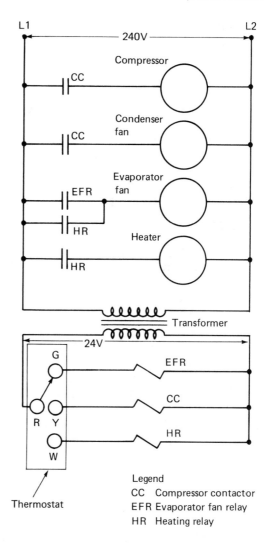

FIGURE 15-3 Ladder diagram for a heat-and-cool installation.

coil is energized, or the heating relay coil. Once the coil of the EFR is energized by having the thermostat turned to make contact with the desired point (G), it closes the points in the relay and the evaporator fan motor starts to move. This means that the low voltage (24 volts) has energized the relay. The relay energizes and closes the EFR contacts located in the high-voltage (240 volts) circuit. If the thermostat is turned to W or the heating position, it will cause the heating relay coil to be energized when the thermostat switch closes and demands heat. The energized heating relay coil causes the HR contacts to close, which in turn places the heating element across the 240-volt line and it begins to heat up. Note that the HR contacts are in parallel with the evaporator fan relay contacts. Thus, the evaporator fan will operate when *either* the heating relay or the evaporator fan relay is energized.

MANUFACTURER'S DIAGRAMS

Figure 15-4 shows how the manufacturer represents the location of the various furnace devices. The solid lines indicate the line voltage to be installed. The dotted lines are the low voltage to be installed when the furnace is put into service.

The motor is four speed. It has various colored leads to represent the speeds. You may have to change the speed of the motor to move the air to a given location. Most motors come from the factory with a medium-high speed selected. The speed is usually easily changed by removing a lead from one point and placing it on another where the proper color is located. In the schematic of Fig. 15-5, the fan motor has white connected to one side of the 120-volt line (neutral) and the red and black are switched by the indoor blower relay to black for the cooling speed and red for the heating speed. It takes a faster fan motor to push the cold air than for hot air because cold air is heavier than hot air.

In Fig. 15-5, the contacts on the thermostat are labeled R, W, Y, G. R and W are used to place the thermostat in the circuit. It can be switched from W to Y manually by moving the Heat–Cool switch on the thermostat to Cool position.

Notice in Fig. 15-5 that the indoor blower relay coil is in the circuit all the time when the AUTO-ON switch on the thermostat is located at the ON position. The schematic also shows the cool position has been selected manually, and the thermostat contacts will complete the circuit when it moves from W1 to Y1.

In Fig. 15-4, note that the low-voltage terminal strip has a T on it. This is the common side of the low voltage from the transformer. In Fig. 15-5, the T is the common side of the low-voltage transformer secondary. In Fig. 15-4, the T terminal is connected to the compressor contactor by a wire run from the terminal to the contactor. Note that the other wire to the contactor runs from Y on the terminal strip. Now go back to Fig. 15-5, where the Y and T terminals are shown as connection points for the compressor contactor. Are you able to relate the schematic to the actual device? The gas valve is wired by having wire T of the terminal strip attached to one side of the solenoid and a wire run from the limit switch to the other side of the solenoid.

Figure 15-6 shows how the wiring diagram comes from the factory. It is usually located inside the cover for the cold-air return. In most instances it is glued to the cover so that it is handy for the person working on the furnace whenever there is a problem after installation.

FIELD WIRING

The installation of a new furnace requires you to follow a factory diagram furnished in a booklet that accompanies the unit. The wiring to be done in the field is represented by the dotted lines in Fig. 15-7. All electrical connections should be made in accordance with the National Electrical Code and any local codes or ordinances that might apply.

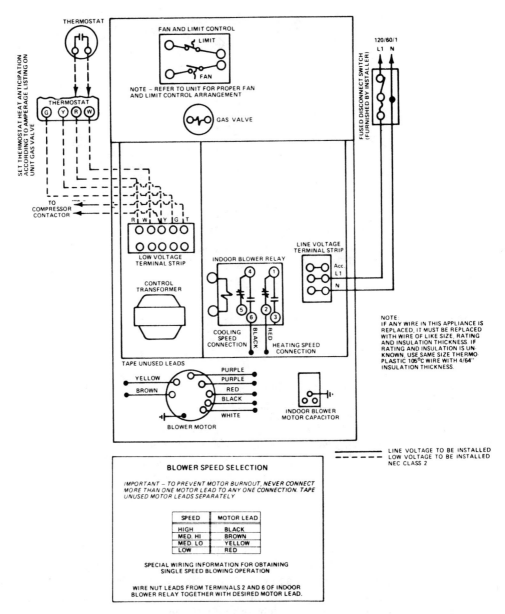

FIGURE 15-4 Manufacturer's diagram for a hot-air installation.

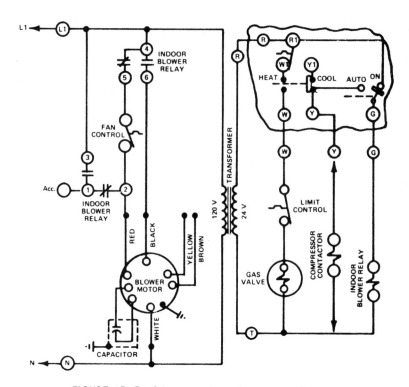

FIGURE 15-5 Schematic for a hot-air installation.

WARNING: The unit cabinet must have an uninterrupted or unbroken electrical ground to minimize personal injury if an electrical fault should occur. This may consist of electrical wire or approved conduit when installed in accordance with existing electrical codes.

LOW-VOLTAGE WIRING

Make the field low-voltage connections at the low-voltage terminal strip shown in Fig. 15-7. Set the thermostat heat anticipator at 0.60 ampere (or whatever is called for by the manufacturer). If additional controls are connected in the thermostat circuit, their amperage draw must be added to this setting. Failure to make the setting will result in improper operation of the thermostat.

With the addition of an automatic vent damper, the anticipater setting would then be 0.12 ampere. As you can see from this and the schematic (see Fig. 14-16), the anticipator resistor is in series with whatever is in the circuit and is to be controlled by the thermostat. The more devices controlled by the thermostat, the more current will be drawn from the transformer to energize them. As the current demand increases, the current through the anticipator is also increased. As you remember

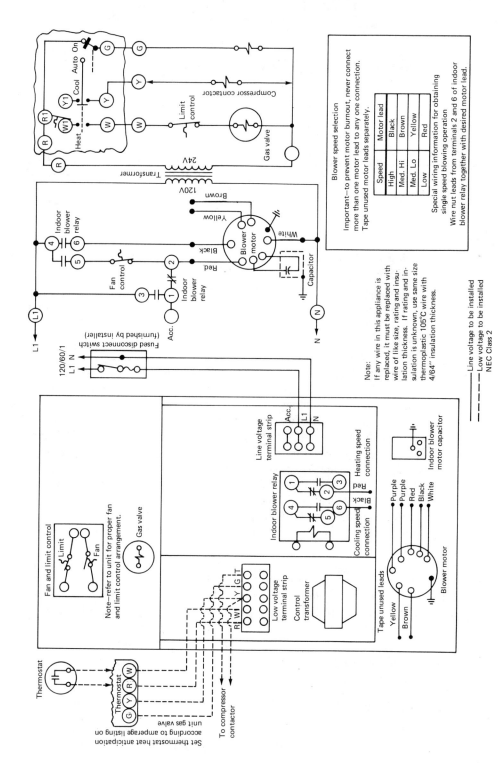

FIGURE 15-6 Complete instruction page packaged with a hot-air furnace.

286 HEATING CIRCUITS

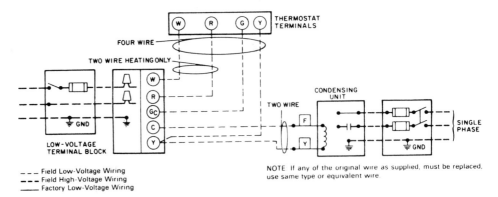

FIGURE 15-7 Heating and cooling application wiring diagram. (Courtesy of Carrier)

from previous chapters, a series circuit has the same current through each component in the circuit.

Thermostat Location. The room thermostat should be located where it will be in the natural circulating path of room air. Avoid locations where the thermostat is exposed to cold-air infiltration, drafts from windows, doors, or other openings leading to the outside, or air currents from warm- or cold-air registers, or to exposure where the natural circulation of the air is cut off, such as behind doors and above or below mantels or shelves. Also keep the thermostat out of direct sunlight.

The thermostat should not be exposed to heat from nearby fireplaces, radios, televisions, lamps, or rays from the sun. Nor should the thermostat be mounted on a wall containing pipes or warm-air ducts or a flue or vent that could affect its operation and prevent it from properly controlling the room temperature. Any hole in the plaster or panel through which the wires pass from the thermostat should be adequately sealed with suitable material to prevent drafts from affecting the thermostat.

Printed Circuit Board Control Center. Newer hot-air furnaces feature printed circuit control. The board shown in Fig. 15-8 is such that it is easy for the technician installing the furnace to hook it up properly the first time. The markings are designed for making it easy to connect the furnace for accessories, if needed. Figures 15-9 and Fig. 15-10 show the factory-furnished schematic. See if you can trace the schematic and locate the various points on the printed circuit boards.

HEAT PUMPS

The heat pump is a heat multiplier. It takes warm air and makes it hot air. This is done by compressing the air and increasing its temperature. Heat pumps have re-

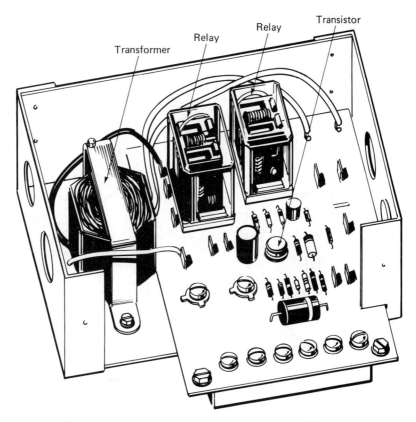

FIGURE 15-8 Printed circuit control center: heat-and-cool models. (Courtesy of Carrier)

ceived more attention since the fuel embargo of 1974. Energy conservation has become a more important concern for everyone. If a device can be made to take heat from the air and heat a home or commercial building, it is very useful to many people.

The heat pump can take the heat generated by a refrigeration unit and use it to heat a house or room. Most of them take the heat from outside the home and move it indoors (see Fig. 15-11). This unit can be used to air condition the house in the summer and heat it in the winter by taking the heat from the outside air and moving it inside.

Operation

On mild temperature heating days, the heat pump handles all heating needs. When the outdoor temperature reaches the balance point of the home, that is, when the heat loss is equal to the heat-pump heating capacity, the two-stage indoor thermostat

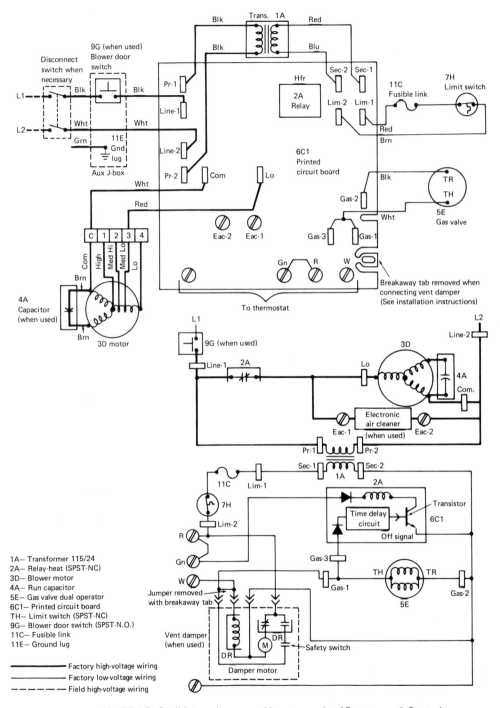

FIGURE 15-9 Wiring diagram. Heating only. (Courtesy of Carrier)

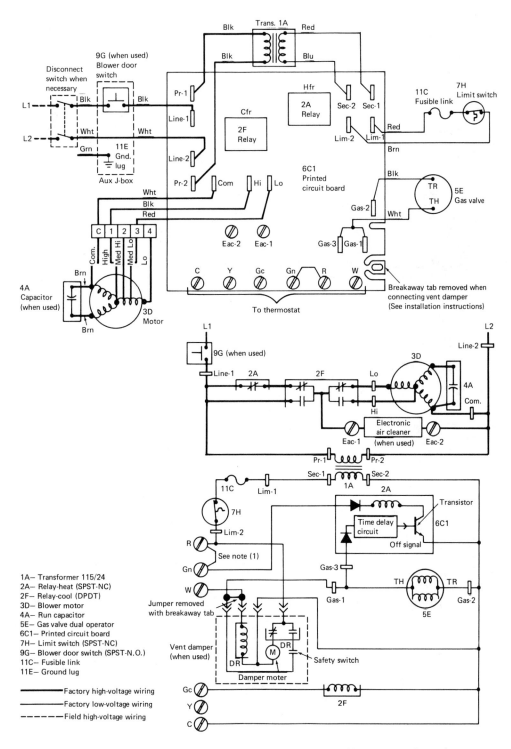

FIGURE 15-10 Wiring diagram. Heat and cool. (Courtesy of Carrier)

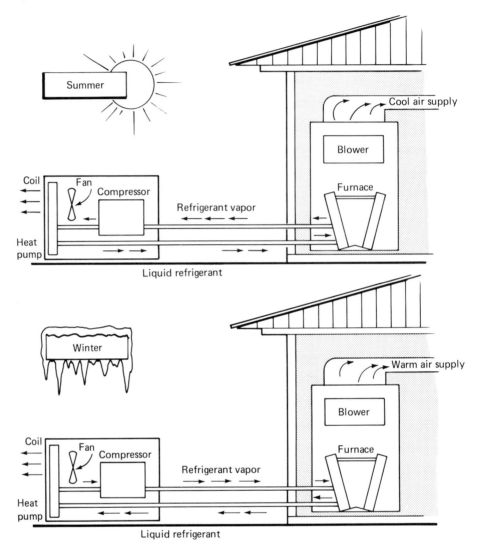

FIGURE 15-11 Basic operation of a heat pump.

activates the furnace (a secondary heat source, in most cases electric heating elements). As soon as the furnace is turned on, a heat relay de-energizes the heat pump.

When the second-stage (furnace) need is satisfied and the plenum temperature has cooled to below 90° and 100°F, the heat-pump relay turns the heat pump back on and controls the conditioned space until the second-stage operation is required again. Figure 15-12 shows the heat-pump unit. The optional electric heat unit shown in Fig. 15-13 is added in geographic locations where needed. This particular unit

FIGURE 15-12 Single-package heat pump.

FIGURE 15-13 Optional electrical heat for a heat pump.

can provide 23,000 to 56,000 Btu's per hour (Btuh) and up to 112,700 Btuh with the addition of electric heat.

If the outdoor temperature drops below the setting of the low-temperature compressor monitor, the control shuts off the heat pump completely and the furnace handles all the heating needs.

During the defrost cycle, the heat pump switches from heating to cooling. To prevent cool air from being circulated in the house when heating is needed, the control automatically turns on the furnace to compensate for the heat-pump defrost cycle (see Fig. 15-14). When supply air temperature climbs above 110° to 120°F,

FIGURE 15-14 Control box for an add-on type of heat pump.

the defrost limit control turns off the furnace and keeps indoor air from getting too warm.

If, after a defrost cycle, the air downstream of the coil gets above 115°F, the closing point of the heat-pump relay, the compressor will stop until the heat exchanger has cooled down to 90° to 100° as it does during normal cycling operation between furnace and heat pump.

During summer cooling, the heat pump works as a normal split system, using the furnace blower as the primary air mover (see Fig. 15-15).

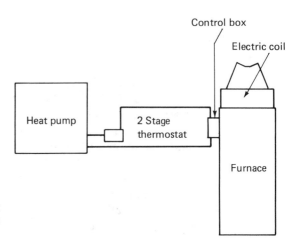

FIGURE 15-15 Heat pump with a two-stage thermostat and control box mounted on the furnace.

In a straight heat pump/supplementary electric heater application, at least one outdoor thermostat is required to cycle the heaters as the outdoor temperature drops. In the system shown here, the indoor thermostat controls the supplemental heat source (furnace). The outdoor thermostat is not required.

Since the furnace is serving as the secondary heat source, the system does not require the home rewiring usually associated with supplemental electric strip heating.

Special Requirements of Heat-Pump Systems

The installation, maintenance, and operating efficiency of the heat-pump system are like those of no other comfort system. A heat-pump system requires the same air quantity for heating and cooling. Because of this, the air-moving capability of an existing furnace is extremely important. It should be carefully checked before a heat pump is added. Heating and load calculations must be accurate. System design and installation must be precise and according to the manufacturer's suggestions.

The air-distribution system and diffuser location are equally important. Supply ducts must be properly sized and insulated. Adequate return air is also required.

Heating supply air is cooler than with other systems. This is quite noticeable to homeowners accustomed to gas or oil heat. This makes diffuser location and system balancing critical.

Heat-Pump Combinations

There are four ways to describe the heat-pump methods of transporting heat into the house:

1. *Air-to-air:* This is the most common method. It is the type of system previously described.

2. *Air-to-water:* This type uses two different types of heat exchangers. Warmed refrigerant flows through pipes to a heat exchanger in the boiler. Heated water flows into radiators located within the heated space.

3. *Water-to-water:* This type uses two water-to-refrigerant heat exchangers. Heat is taken from the water source (well water, lakes, or the sea) and is passed on by the refrigerant to the water used for heating. The reverse takes place in the cooling system.

4. *Water-to-air:* Well water furnishes the heat. This warms the refrigerant in the heat-exchanger coil. The refrigerant, compressed, flows to the top of the unit, where a fan blows air past the heat exchanger.

Each type of heat pump has its advantages and disadvantages. Each needs to be properly controlled. This is where the electrical connections and controls are used to do the job properly. Before attempting to work on this type of equipment, make sure you have a complete schematic of the electrical wiring and know all the component parts of the system.

HIGH-EFFICIENCY FURNACES

Newer furnaces (since 1981) have been designed with efficiencies of up to 97%, as compared to older types with efficiencies in the 60% range. The Lennox^R Pulse^{TM} is one example of the types available.

The G14 series pulse combustion up-flow gas furnace provides efficiency of up to 97%. Eight models for natural gas and LPG are available with input capacities of 40,000, 60,000, 80,000, and 100,000 Btuh. The units operate on the pulse-combustion principle and do not require a pilot burner, main burners, conventional flue, or chimney. Compact, standard-sized cabinet design, with side or bottom return air entry, permits installation in a basement, utility room, or closet. Evaporator coils may be added, as well as electronic air cleaners and power humidifiers (see Fig. 15-16).

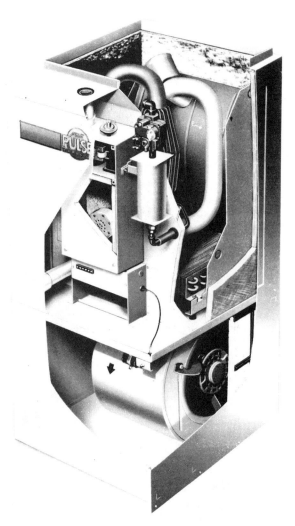

FIGURE 15-16 Lennox Pulse Furnace. (Courtesy of Lennox)

Operation

The high-efficiency furnaces achieve that level of fuel conversion by using a unique heat-exchanger design. It features a finned cast-iron combustion chamber, temperature-resistant steel tailpipe, aluminized steel exhaust decoupler section, and a finned stainless-steel tube condenser coil similar to an air-conditioner coil. Moisture, in the products of combustion, is condensed in the coil, thus wringing almost every usable Btu out of the gas. Since most of the combustion heat is utilized in the heat transfer from the coil, flue vent temperatures are as low as 100°F to 130°F, allowing for the use of 2-inch diameter polyvinyl chloride (PVC) pipe. The furnace is vented through a side wall or roof or to the top of an existing chimney with up to 25 feet of PVC pipe and four 90° elbows. Condensate created in the coil may be disposed of in an indoor drain (see Fig. 15-17). The condensate is not harmful to standard household plumbing and can be drained into city sewers and septic tanks without damage.

The furnace has no pilot light or burners. An automotive-type spark plug is used for ignition on the initial cycle only, saving gas and electrical energy. Due to the pulse-combustion principle, the use of atmospheric gas burners is eliminated, with the combustion process confined to the heat-exchanger combustion chamber. The sealed combustion system virtually eliminates the loss of conditioned air due to combustion and stack dilution. Combustion air from the outside is piped to the furnace with the same type of PVC pipe used for exhaust gases.

Electrical Controls

The furnace is equipped with a standard-type redundant gas valve in series with a gas expansion tank, gas intake flapper valve, and air intake flapper valve. Also factory installed are a purge blower, spark plug igniter and flame sensor with solid-state control circuit board. The standard equipment includes a fan and limit control, a 30-VA transformer, blower cooling relay, flexible gas line connector, and four

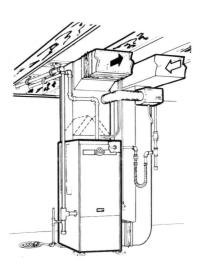

FIGURE 15-17 Basement installation of the Pulse® with cooling coil and automatic humidifier. Note the floor drain for condensate. (Courtesy of Lennox)

296 HEATING CIRCUITS

isolation mounting pads, as well as a base insulation pad, condensate drip leg, and cleanable air filter. Flue vent/air intake line, roof or wall termination installation kits, LPG conversion kits, and thermostat are available as accessories and must be ordered extra, or you can use the existing one when replacing a unit.

The printed circuit board is replaceable as a unit when there is a malfunction of one of the components. It uses a multivibrator transistorized circuit to generate the high voltages needed for the spark plug. The spark plug gets very little use except to start the combustion process. It has a long life expectancy. Spark gap is 0.115 inch and the ground electrode is adjusted to 45° (see Fig. 15-20).

Sequence of Operation

On a demand for heat, the room thermostat initiates the purge blower operation for a prepurge cycle of 34 seconds, followed by energizing of the ignition and opening of the gas valve. As ignition occurs, the flame sensor senses proof of ignition and de-energizes the spark igniter and purge blower (see Fig. 15-18). The furnace

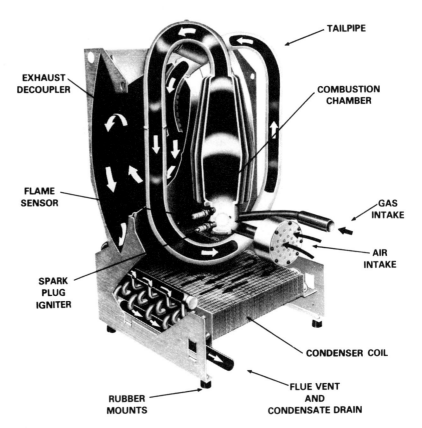

FIGURE 15-18 Cutaway view of the Pulse ™ furnace combustion chamber. (Courtesy of Lennox)

ELECTRICAL

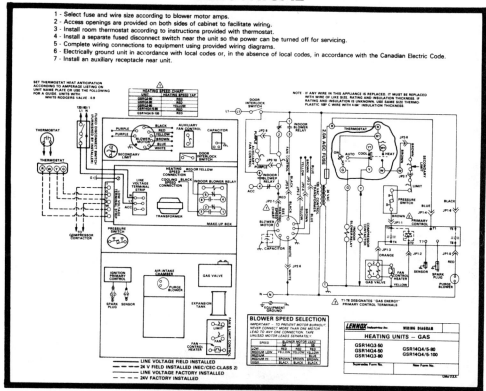

START-UP/ADJUSTMENTS

START-UP
This unit is equipped with a direct spark ignition system with flame rectification. Once combustion has started, the purge blower and spark ignitor are turned off. To place furnace in operation:

1 - With thermostat set below room temperature make sure gas valve knob is in off position and wait 5 minutes.
2 - Turn manual knob of gas valve counterclockwise to ON position. Turn power on and set thermostat above room temperature. Unit will go into prepurge for approximately 30 seconds and then ignite.
3 - If the unit does not light on the first attempt, it will attempt four more ignitions before locking out.
4 - If lockout occurs, turn thermostat off and then back on.

To shut off furnace:
1 - Set thermostat to lowest temperature and turn power supply to furnace off.
2 - Turn manual knob of gas valve off.

FAILURE TO OPERATE
If unit fails to operate, check the following:

1 - Is thermostat calling for heat?
2 - Is main disconnect switch closed?
3 - Is there a blown fuse?
4 - Is filter dirty or plugged? Dirty or plugged filters will cause unit to go off on limit control.
5 - Is gas turned on at meter?
6 - Is manual main shut-off valve open?
7 - Is internal manual shut-off open?
8 - Are intake and exhaust pipes clogged?
9 - Is primary control locked out? (Turn thermostat off and then back on.)
10 - Is unit locked out on secondary limit? (Secondary limit is manually reset.)

GSR14 SERIES UNITS

LENNOX Industries Inc.

FIGURE 15-19 Electrical, start-up adjustments, maintenance, and repair list for the GSR-14Q furnaces. (Courtesy of Lennox)

ADJUSTMENTS (CONT.)

GAS FLOW

To check proper gas flow to combustion chamber, determine BTU input from the appliance rating plate. Divide this input rating by the BTU per cubic foot of available gas. Result is the number of cubic feet per hour required. Determine the flow of gas through gas meter for 2 minutes and multiply by 30 to get the hourly flow of gas to burner.

GAS PRESSURE

1 - Check gas line pressure with unit firing at maximum rate. Normal natural gas inlet line pressure should be 7.0 in. (178 mm) w.c. Normal line pressure for LP gas is 11 in. (280 mm) w.c.

IMPORTANT - Minimum gas supply pressure is listed on unit rating plate for normal input. Operation below minimum pressure may cause nuisance lockouts.

2 - After line pressure is checked and adjusted, check regulator pressure. Correct manifold pressure (unit running) is specified on nameplate. To measure, connect gauge to pressure tap in elbow below expansion tank.

HEAT ANTICIPATOR SETTINGS

Units with White Rodgers gas valves — 0.9

FAN/LIMIT CONTROL

Limit Control — Factory set: No adjustment necessary
Fan Control — Factory set: ON — No adjustment necessary
OFF — 90°F (32°C)

TEMPERATURE RISE AND EXTERNAL STATIC PRESSURE

Check temperature rise and external static pressure. If necessary, adjust blower speed to maintain temperature rise and external static pressure within range shown on unit rating plate.

ELECTRICAL

1 - Check all wiring for loose connections.
2 - Check for correct voltage at unit (unit operating).
3 - Check amp-draw on blower motor.
   Motor Nameplate_____ Actual_____

NOTE - Do not secure electrical conduit directly to duct work or structure.

BLOWER SPEEDS

Multi-tap drive motors are wired for different heating and cooling speeds. Speed may be changed by simply interchanging motor connections at indoor blower relay and fan control. Refer to speed selection chart on unit wiring diagram.

CAUTION - To prevent motor burnout, never connect more than one (1) motor lead to any one connection. Tape unused motor leads separately.

MAINTENANCE

NOTE - Disconnect power before servicing.

ANNUAL MAINTENANCE

At the beginning of each heating season, system should be checked by qualified serviceman as follows:

A - Blower

1 - Check and clean blower wheel.
2 - Check motor lubrication.
   Always relubricate motor according to manufacturer's lubrication instructions on each motor. If no instructions are provided, use the following as a guide:
   a - Motors Without Oiling Ports — Prelubricated and sealed. No further lubrication required.
   b - Direct Drive Motors with Oiling Ports — Prelubricated for an extended period of operation. For extended bearing life, relubricate with a few drops of SAE No. 10 non-detergent oil once every two years. It may be necessary to remove blower assembly for access to oiling ports.

B - Electrical

1 - Check all wiring for loose connections.
2 - check for correct voltage at unit (unit operating).
3 - Check amp-draw on blower motor.
   Motor nameplate _____ Actual _____
4 - Check to see that heat tape (if applicable) is operating.

C - Filters

1 - Filters must be cleaned or replaced when dirty to assure proper furnace operation.
2 - Reusable foam filters supplied with GSR14 can be washed with water and mild detergent. They should be sprayed with filter handicoater when dry prior to reinstallation. Filter handicoater is RP products coating No. 481 and is available as Lennox part No. P-8-5069.
3 - If replacement is necessary, order Lennox part No. P-8-7831 for 20 X 25 inch (508 X 635 mm) filter.

D - Intake and Exhaust Lines

Check intake and exhaust PVC lines and all connections for tightness and make sure there is no blockage. Also check condensate line for free flow during operation.

E - Insulation

Outdoor piping insulation should be inspected yearly for deterioration. If necessary, replace with same materials.

FIGURE 15-19 (B) *(continued)*

HIGH-EFFICIENCY FURNACES

REPAIR PARTS LIST

The following repair parts are available through independent Lennox dealers. When ordering parts, include the complete furnace model number listed on unit rating plate. Example: GSR14Q3-50-1.

CABINET PARTS
Blower access panel
Control access panel
Upper vestibule panel
Lower vestibule panel
Control box cover

CONTROL PANEL PARTS
Transformer
Indoor blower relay
Low voltage terminal strip
High voltage terminal strip

BLOWER PARTS
Blower wheel
Motor
Motor mounting frame
Motor capacitor
Blower housing cut-off plate
Blower housing

HEATING PARTS
Heat exchanger assembly
Gas orifice
Gas valve
Gas decoupler
Gas flapper valve
Purge blower
Air intake flapper valve
Primary control board
Ignition lead
Flame sensor lead
Flame sensor
Primary fan and limit control
Secondary limit control
Auxiliary fan control
Differential pressure switch
Door interlock switch
Air filter

FIGURE 15-19 (C) (*continued*)

blower operation is initiated 30 to 45 seconds after combustion ignition. When the thermostat is satisfied, the gas valve closes and the purge blower is re-energized for a post-purge cycle of 34 seconds. The furnace blower remains in operation until the preset temperature setting (90°F) of the fan control is reached. Should the loss of flame occur before the thermostat is satisfied, flame sensor controls will initiate three to five attempts at reignition before locking out the unit operation. In addition, loss of either combustion intake air or flue exhaust will automatically shut down the system.

Combustion Process

The process of pulse combustion begins as gas and air are introduced into the sealed combustion chamber with the spark plug igniter. Spark from the plug ignites the gas/air mixture, which in turn causes a positive pressure buildup that closes the gas and air inlets. This pressure relieves itself by forcing the products of combustion out of the combustion chamber through the tailpipe into the heat-exchanger exhaust decoupler and on into the heat-exchanger coil. As the combustion chamber empties, its pressure becomes negative, drawing in air and gas for ignition of the next pulse. At the same instant, part of the pressure pulse is reflected back from the tailpipe at the top of the combustion chamber. The flame remnants of the previous pulse of combustion ignite the new gas/air mixture in the chamber, continuing the cycle.

TROUBLESHOOTING

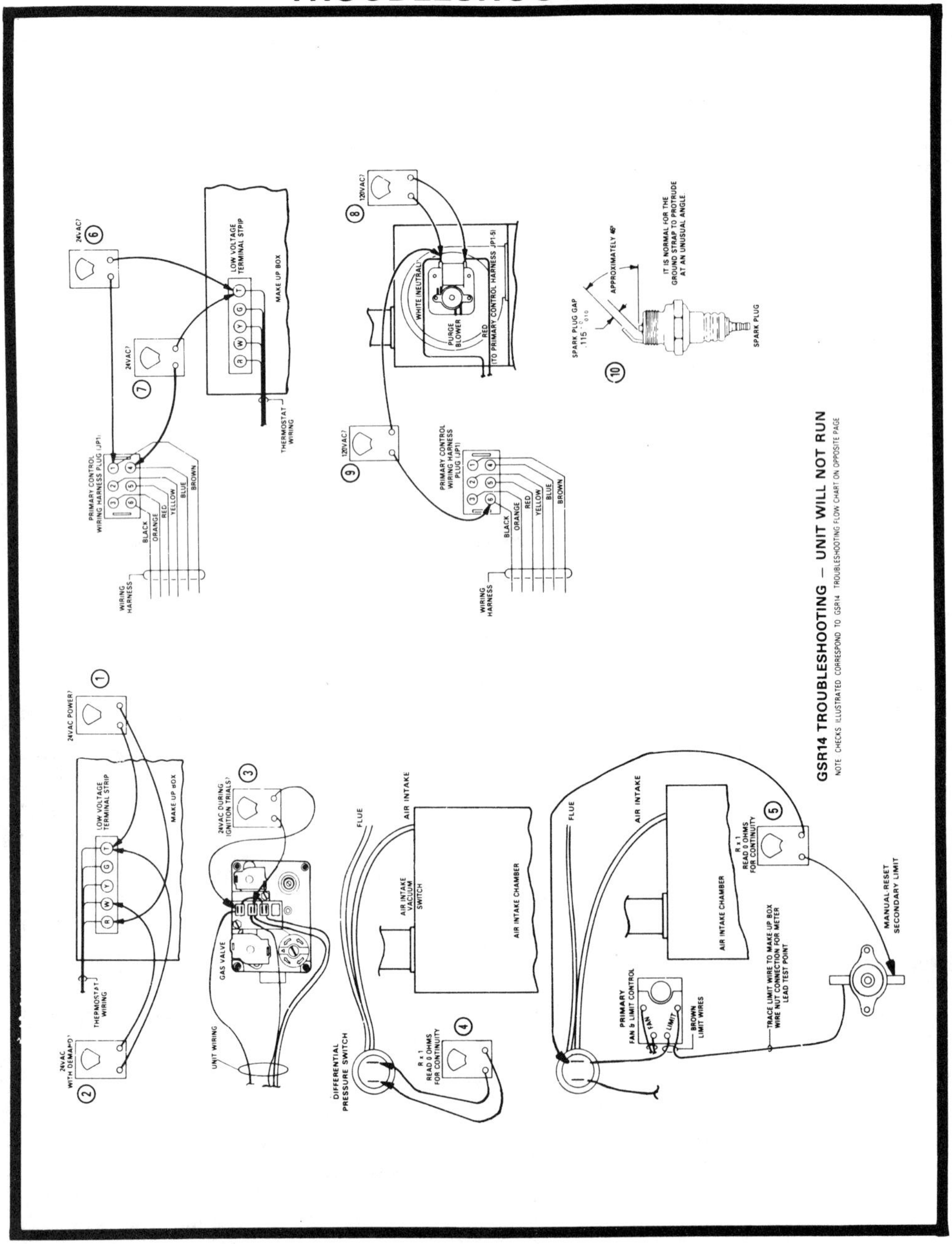

FIGURE 15-20 Troubleshooting the GSR-14Q furnace with a meter. (Courtesy of Lennox)

HIGH-EFFICIENCY FURNACES

TROUBLESHOOTING (CONT.)

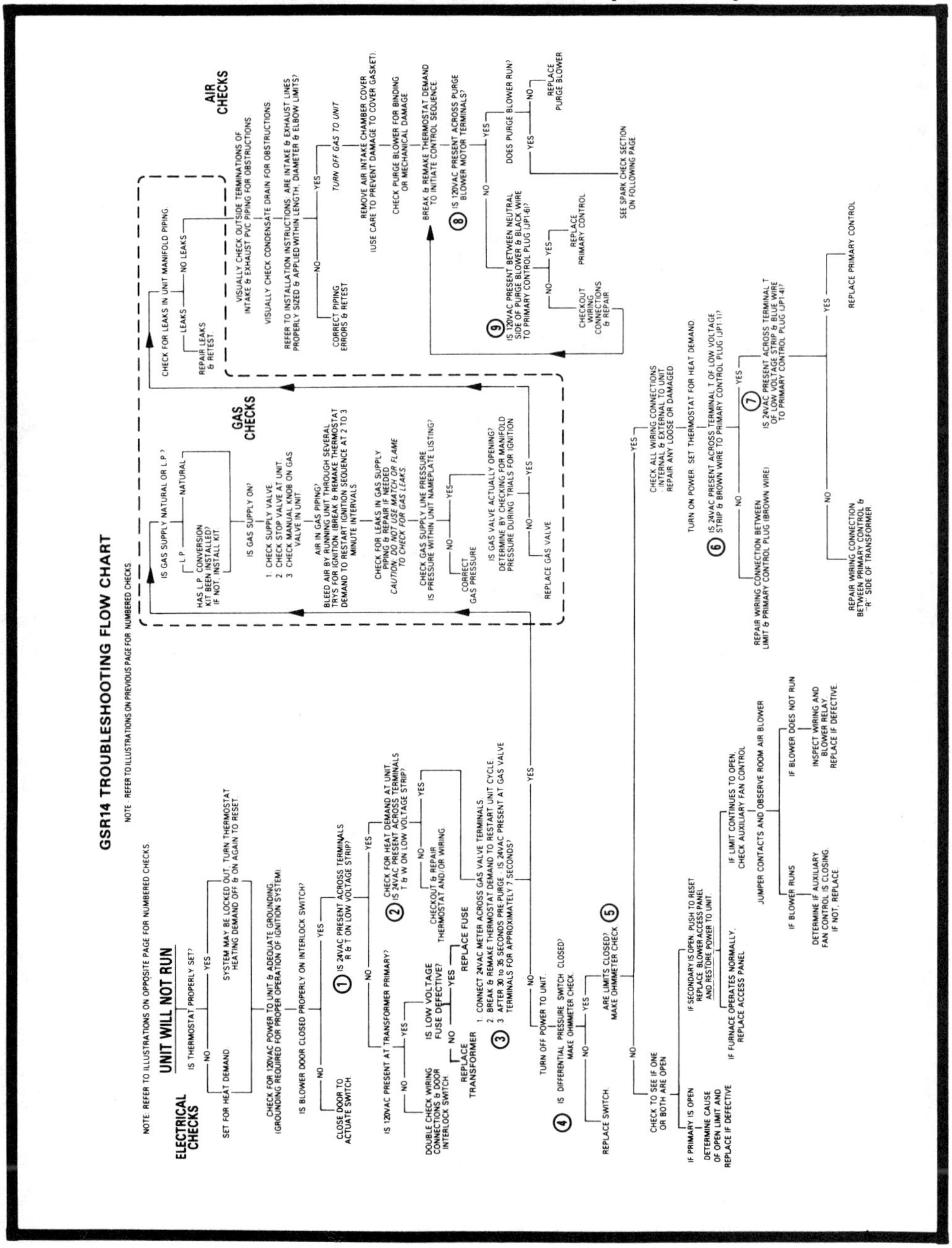

FIGURE 15-20 (B) (continued)

TROUBLESHOOTING (CONT.)

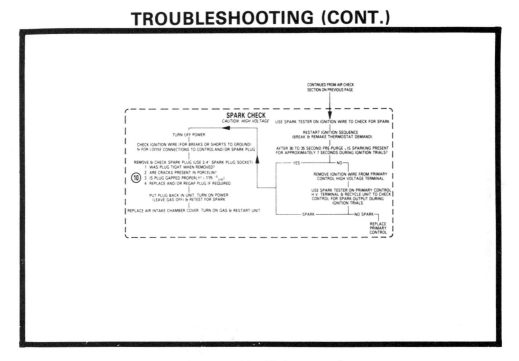

FIGURE 15-20 (C) (*continued*)

Once combustion is started, it feeds on itself, allowing the purge blower and spark igniter to be turned off. Each pulse of gas/air mixture is ignited at the rate of 60 to 70 times per second, producing one-fourth to one-half Btu per pulse of combustion. Almost complete combustion occurs with each pulse. The force of these series of ignitions creates great turbulence, which forces the products of combustion through the entire heat-exchanger assembly, resulting in maximum heat transfer (see Fig. 15-18).

Start-up procedures for the GSR-14Q series of Pulse furnaces, as well as maintenance and repair parts, are shown in Fig. 15-19.

TROUBLESHOOTING THE PULSE™ FURNACE

Troubleshooting procedures for the Pulse furnaces are shown in Fig. 15-20. Figure 15-21 shows the circuitry for the G-14Q series of furnaces. Note the difference in the electrical circuitry for the G-14 and GSR-14. Blower speed color-coded wires are also indicated for the different units. The 40, 60, 80, and 100 after the G-14Q indicates whether it is a 40,000, 60,000, 80,000, or 100,000 Btuh unit. Thermostat

ELECTRICAL

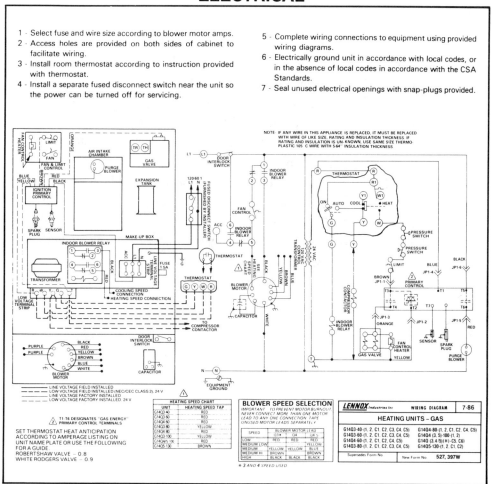

FIGURE 15-21 Electrical wiring for the G-14Q series of furnaces. (Courtesy of Lennox)

heat anticipation is also given for the Robertshaw valve and the Rodgers valve. This type of electrical diagram is usually glued to the cabinet so that it is with the unit whenever there is need for troubleshooting.

The troubleshooting flow chart is typical of those furnished with newer equipment in the technical manuals furnished the dealers who provide the service. After locating the exact symptoms, check with the other part of Fig. 15-20 to find how to use the multimeter to check out all the circuitry to see if the exact cause of the problem can be determined.

REVIEW QUESTIONS

1. What are the parts of a basic electric heating system?
2. What is a ladder diagram? Why is it useful?
3. What is the source of schematics for specific types of equipment?
4. What is the meaning of the term field wiring?
5. What is the usual low voltage for a heating system?
6. Where should the thermostat be located in a home?
7. What is the advantage of a printed circuit board for a furnace?
8. How is a heat pump different from a furnace?
9. What does a heat pump do that a furnace cannot do?
10. How can a refrigerant be used to generate heat?
11. List four basic types of heat pumps.
12. Why are supplemental heaters sometimes needed on heat pumps?
13. How is the gas/air mixture ignited in a Pulse furnace?
14. Why are these new furnaces so efficient?
15. Where is the electronics package located in the Pulse furnace?

PERFORMANCE OBJECTIVES

Know what is meant by *modes of cooling*.
Know where the temperature sensor is located on an air conditioner.
Know how to adjust the fan speed on an air conditioner.
Know what an auxiliary winding does.
Know the purpose of a run capacitor in the compressor circuit.
Know how to read a ladder diagram.
Know the function of the AUTO and ON positions of a thermostat.
Know how the overload protector works in an air conditioner compressor motor.
Be able to troubleshoot an air conditioner using a troubleshooting chart.

CHAPTER 16

Air-Conditioning Circuits

Circuits for air-conditioning units vary slightly with the manufacturer. The best source for schematics of the electrical system is the manufacturer's service manuals and bulletins. In this chapter we will be concerned primarily with the circuits of domestic (home) air-conditioning equipment.

BASIC AIR-CONDITIONING UNIT

The basic or simple air-conditioning unit consists of a plug, a thermostat, overload protector, compressor, condenser, evaporator, capacitors for the motors, and some kind of switching for control of the fan motor.

The shaft on the fan motor usually extends from both ends of the motor. A fan blade is attached to one end to cool the condenser, and another fan blade is attached to the other end of the motor to cause air to flow over the evaporator and cool the room. This way only one motor is used. This is the type of arrangement most often encountered in the window air conditioner.

The air-conditioning unit that fits into the window (see Fig. 16-1) usually has at least two speeds. The fan is used in some cases without the compressor running. Thus, there are *three modes of cooling:* (1) fan only, (2) fan at high speed and cooling furnished by the compressor, and (3) fan at low speed and cooling furnished by the compressor. This basic window unit consists of the compressor, the fan motor, the thermostat, and push buttons for the selection of the type of cooling desired (see Fig. 16-2).

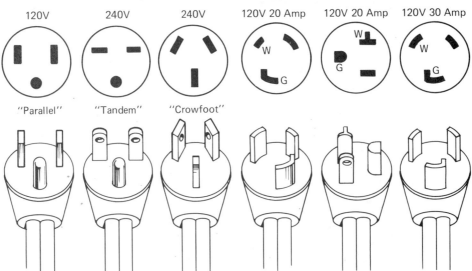

FIGURE 16-1 (A) Window air conditioner. (B) Line cords and plugs for various voltages used on air-conditioner units.

Push-button Switches

The push-button switches that operate the window unit usually resemble those in Fig. 16-3. If the unit has a two-speed fan and a compressor that operates in conjunction with the high- and low-speed fan positions, it will have an OFF button that disengages the push buttons that are depressed and thereby opens the circuit and turns off the power to both compressor and fan.

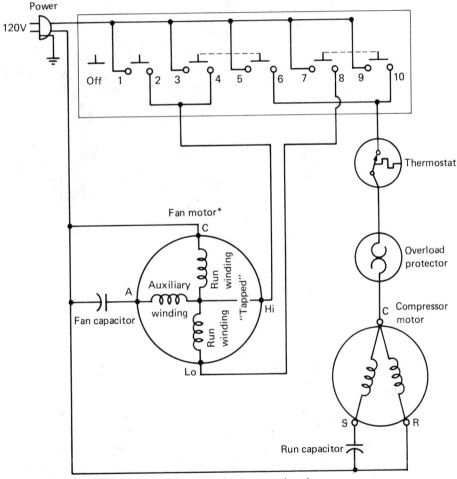

*Extended shaft fan motor. Fan blades on each end.
One fan for evaporator and one for condenser.

FIGURE 16-2 (A) Schematic for a two-speed, push-button-controlled window unit. (B) Location of component parts of a window unit.

310 AIR-CONDITIONING CIRCUITS

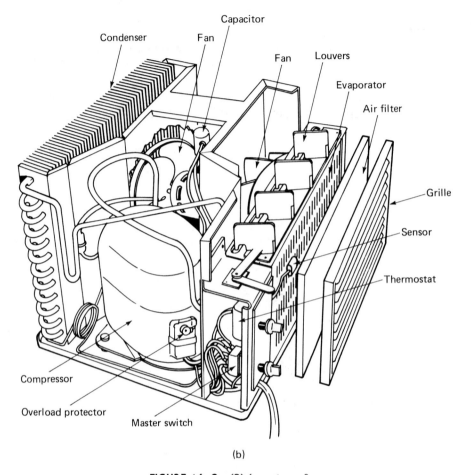

(b)

FIGURE 16-2 (B) (*continued*)

Thermostat

The thermostat is variable by twisting a knob. This moves the bimetallic strips closer together to provide the proper operation of the thermostat in relation to the room temperature. Thermostats can have a number of design configurations, depending on the manufacturer. The knobs differ slightly, but basically the thermostat is the same for all window units. The temperature sensor located on the inlet side of the evaporator senses the air returned to the air conditioner from the room being conditioned.

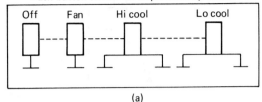

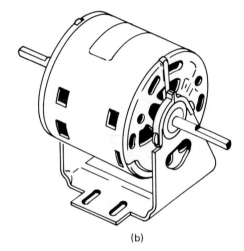

FIGURE 16-3 (A) Push-button switches used in a window unit for fan and compressor control. (B) Fan with double shaft. Used in window units with a blade on each end.

Fan

The fan is two speed. It has an auxiliary winding with a capacitor in series with it. The run windings are arranged so that when one of them is in the circuit it runs at high speed. When the two run windings are placed in series, it slows down and the fan speed is LO. This permanent split-capacitor (PSC) motor is commonly used in two- or three-speed configurations for fans.

SCHEMATICS

To be able to troubleshoot this type of window unit, it is necessary to know how each control functions and what will happen when each malfunctions. The best way to understand the unit is to look at the schematic in Fig. 16-2, and trace it out.

Tracing it out will result in Fig. 16-4. In Fig. 16-4A, you can see how one side of the power source is connected to the fan motor C and run capacitor. Then it goes to the run capacitor on the compressor and the R connection of the compressor.

312 AIR-CONDITIONING CIRCUITS

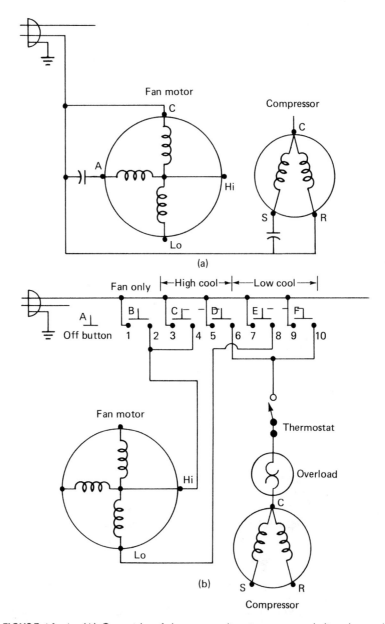

FIGURE 16-4 (A) One side of the power line is connected directly to all components in the unit. (B) The other side of the power line is switched and becomes the controlled power line. (C) Schematic for FAN ONLY operation of the unit. (D) Schematic for HI COOL operation of the unit. (E) Schematic for LO COOL operation of the unit.

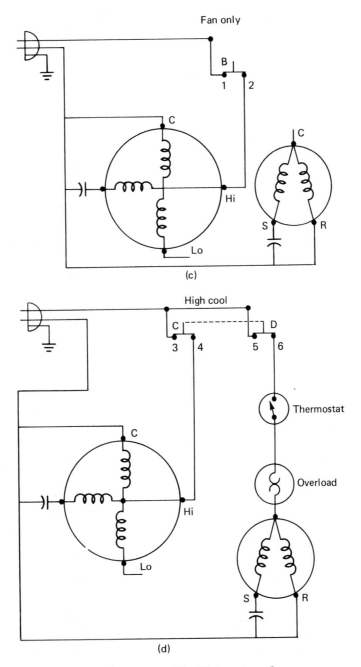

FIGURE 16-4 (C), (D) (continued)

314 AIR-CONDITIONING CIRCUITS

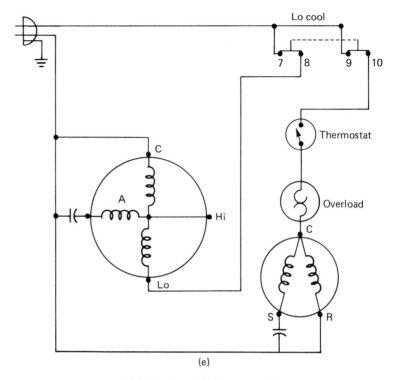

FIGURE 16-4 (E) (*continued*)

This completes the line for one side of the power. But, for anything to work electrically, two sides of the power line must be taken into consideration. It takes both sides to cause the circuit to be complete. Now let's look at the other line as seen in Fig. 16-4B.

All the controls are in this other line. The push-button switch has an OFF button, a FAN ONLY button, a HI COOL button, and a LO COOL button. Each of these has a definite circuit that must be traced out to aid in troubleshooting later when possible switching problems develop.

Figure 16-4C shows the FAN ONLY circuit. Push-button B is depressed. It makes contact with 1 and 2 on the switch block. This completes the other side of the power line through the switch to the high connection of the fan motor. Thus, power is now available to cause the HI WINDING of the fan motor to be energized. The AUX WINDING is also completed to the power line at this time, and the fan motor runs. However, this is the only thing operational in the window unit at this time. The fan is sometimes needed to circulate air in the room when a number of people are assembled. The fan only is often all that is needed to keep the room cool. Keep in mind that the thermostat is not in the circuit, so the fan will run until it is turned off by depressing the OFF button.

Figure 16-4D shows how the compressor is operated with a demand for HI COOL. The HI COOL push button is depressed and it completes the circuit from 3 and 4 to the HI side of the fan motor. The fan runs. HI COOL also is so arranged that the push button will complete the circuit from 5 to 6. The 5 and 6 combination completes the circuit to the thermostat and, if it is closed, on to the overload and to the compressor motor. The compressor will now run while the fan is also running. Once the thermostat opens (the room has cooled down to where the thermostat has been adjusted), the compressor is turned off. However, as long as the HI COOL button is depressed, the fan motor will continue to operate. The compressor is cycled on and off by the demands of the thermostat. The fan will continue to run until the OFF button is depressed to mechanically push up the HI COOL switch.

Figure 16-4E shows the completed circuit for the LO COOL position of the push buttons. Note how terminals 7 and 8 are connected to complete the circuit to the LO position of the fan motor. This means that the two main run windings of the fan motor are now in series. This causes the fan motor to run at a slower speed than when only one winding is in the circuit.

Terminals 9 and 10 are used to complete the circuit to cause the compressor to operate. Here, again, the fan runs at LO speed all the time, whether the compressor operates or not. The compressor operates at the command of the thermostat.

To troubleshoot the unit, you should understand how it operates when all systems are functional. Then the symptoms can be traced to either the refrigerant system or electrical system. Here we are primarily interested in the problems associated with the electrical system. Problems are shown in Table 16-1, which covers both systems. Note how many of the problems are electrical.

TABLE 16-1 Troubleshooting Hermetic Compressors Used in Air Conditioners

Problem	Probable Cause	Remedy
Compressor won't start. There is no hum.	1. Open line circuit.	1. Check the wiring, fuses, and receptacle.
	2. Protector open.	2. Check current drawn from the line after waiting for the reset to cool down.
	3. Contacts open on control relay.	3. Check the control and check the pressure readings.
	4. Open circuit in the motor stator.	4. Replace the stator or the whole compressor.
Compressor hums intermittently but will not start. However, it will cycle with the protector.	1. Incorrectly wired.	1. Check the wiring diagram and the actual wiring.
	2. Low line voltage.	2. Check the line voltage. Find the cause of the dropping voltage and correct.
	3. Start capacitor could be open.	3. Replace with one known to be good.

(continued)

TABLE 16-1 (continued)

Problem	Probable Cause	Remedy
	4. Contacts on relay do not close.	4. Check by manually operating the relay. Replace if defective.
	5. Start winding could be open.	5. Check the stator leads. Replace the compressor if the leads check out and are all right.
	6. Stator winding could be grounded. (This usually blows the fuse.)	6. Check stator leads. Replace compressor if the leads are OK.
	7. Discharge pressure could be too high.	7. Remove the cause of excessive pressure. Discharge shutoff and receiver valves should be open.
	8. Discharge pressure too high.	8. Compressor too tight.
	9. Start capacitor weak.	9. Replace the start capacitor.
Compressor starts, but the motor will not speed up enough to have start winding drop out of the circuit.	1. Low line voltage.	1. Increase the voltage.
	2. Compressor incorrectly wired.	2. Rewire according to wiring diagram.
	3. Defective relay.	3. Check operation. If defective, replace.
	4. Run capacitor shorted.	4. Disconnect the run capacitor and check for short.
	5. Start and run windings shorted.	5. Check winding resistance. If incorrect, replace the compressor.
	6. Start capacitor weak.	6. Check the capacitors. Replace those found to be defective.
	7. Discharge pressure too high.	7. Check the discharge shutoff valves. Check the pressure.
	8. Tight compressor.	8. Check the compressor oil level. Check for binding. Replace if necessary.
Compressor starts and runs, but it cycles on the protector.	1. Low line voltage.	1. Increase the voltage.
	2. Too much current being drawn through the protector.	2. Check to see if the fans or pumps are wired to the wrong connector.
	3. Suction pressure is too high.	3. Check the compressor. See if it is the right size for the job.
	4. Discharge pressure is too high.	4. Check the ventilation of the compressor.

TABLE 16-1 (continued)

Problem	Probable Cause	Remedy
		Check for overcharge. Also check for obstructions to air flow or refrigerant flow.
	5. Protector is weak.	5. Check the current. Replace the protector if it is not clicking out at the right point.
	6. Defective run capacitor.	6. Check the capacitance. Replace if found to be defective.
	7. Stator partially shorted or grounded.	7. Check the resistance for a short to the frame. Replace if found shorted to the frame.
	8. Insufficient motor cooling.	8. Correct air flow.
	9. Compressor tight.	9. Check oil level. Check the cause of the binding.
	10. Three-phase line unbalanced.	10. Check each leg or phase. Correct if the voltages are not the same between legs.
	11. Discharge valve leaks or is damaged.	11. Replace the valve plate.
Start capacitors burn out.	1. Short cycling.	1. Reduce the number of starts. They should not exceed 20 per hour.
	2. Start winding left in circuit too long.	2. Reduce the starting load. Install a crankcase pressure-limit valve. Increase the low voltage if this is found to be the condition. Replace the relay if it is found to be defective.
	3. Relay contacts are sticking.	3. Clean the relay contacts. Or replace the relay.
	4. Wrong relay or wrong relay setting.	4. Replace the relay.
	5. Wrong capacitor.	5. Check the specs for correct-sized capacitor. Be sure the MFD and the WVDC are correct for this compressor.

(*continued*)

318 AIR-CONDITIONING CIRCUITS

TABLE 16-1
(continued)

Problem	Probable Cause	Remedy
	6. Working voltage of the capacitor is too low.	6. Replace with a capacitor of correct voltage rating.
	7. Water shorts out the terminals of the capacitor.	7. Place the capacitor so the terminals will not get wet.
Run capacitors burn out. They spew their contents over the surfaces of anything nearby. This problem can usually be identified with a visual inspection.	1. Line voltage too high.	1. Reduce the line voltage. Should not be over 10% of motor rating.
	2. Light load with a high line voltage.	2. Reduce voltage if not within the 10% overage limit.
	3. Voltage rating of the capacitor too low.	3. Replace with capacitors of the correct WVDC.
	4. Capacitor terminals shorted by water.	4. Place the capacitor so the terminals will not get wet.
Relays burn out.	1. Low line voltage causes high current drain.	1. Increase the voltage to within the 10% limit.
	2. High line voltage.	2. Reduce voltage to within 10% of the motor rating.
	3. Wrong-sized capacitor.	3. Use the correct-sized capacitor. The proper MFD rating should be installed.
	4. Short cycling.	4. Decrease the number of starts per hour.
	5. Relay vibrates.	5. Make sure you mount the relay rigidly.
	6. Wrong relay.	6. Use the recommended relay for the compressor motor.

There are problems of a general nature. They are identified with hermetic-type compressors.

Window units with three-speed fans are also available. All you do is place windings in series with one another to reduce the fan speed. The push buttons do the same things as just explained for the two-speed unit. In larger compressors for 18,000 Btu units, a start potential relay is used to switch in the start winding. A start capacitor is placed in and out of the circuit as required, and the run capacitor stays in the circuit. This type of arrangement is discussed in detail in Chapter 17 for refrigerators.

LADDER DIAGRAMS

Ladder diagrams make it easier to visualize the circuitry. The two power lines are drawn parallel with one another and the components are connected between the two lines to complete the various system circuits (see Fig. 16-5).

Another air-conditioning unit is shown in Fig. 16-5. It utilizes 220 to 240 volts and is wired into the line permanently. This type of unit usually becomes part of a

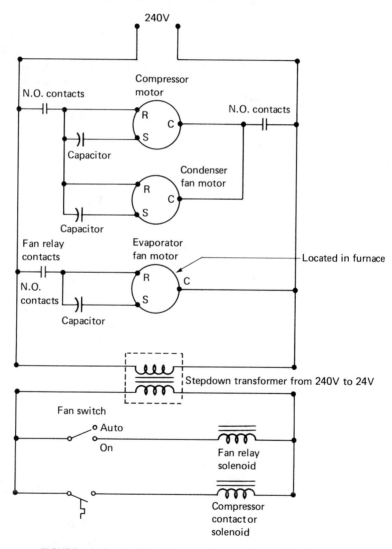

FIGURE 16-5 Ladder diagram for an air-conditioning unit.

hot-air type of furnace and provides cooling in the summer and heating in the winter. The evaporator is located in the plenum of the furnace, and the blower motor is used to move air over the evaporator during the air-conditioning season and move the hot air in the plenum from the furnace during the winter.

Thermostat

The thermostat is in the COOL position. This means that when the room reaches a high temperature the thermostat closes. Once the thermostat closes, the circuit to the compressor contactor (CC) relay is completed. The contacts CC are closed and complete the circuit to the compressor and condenser. The compressor begins to pump the refrigerant, and the fan starts to cool the condenser. Note how the condenser fan motor is connected in parallel with the compressor. The condenser is on when the compressor is on and off when the compressor is off. Both are controlled by the CC contacts. Opening of the thermostat causes the compressor to stop running, and the fan motor for the condenser also stops until needed again when the thermostat switch closes.

Fan Switch

The fan switch has two positions: AUTO and ON. When it is in the ON position, it means the fan relay solenoid is energized and the fan runs continually until the switch is returned to AUTO. When the fan switch is on AUTO, it works in conjunction with the thermostat. When the thermostat closes, the fan switch closes. Thus, it starts the fan only when the compressor is operating. However, if you wish to move the air in the room without the benefit of the cooling action of the compressor, you may do so by manually moving the switch to the ON position.

Note that the fan relay contacts have to be closed in order for the evaporator fan motor to operate. These relay contacts close when the fan relay solenoid energizes. This solenoid and the compressor contactor solenoid are both powered from a 24-volt transformer that provides the low voltage necessary for their operation. This low-voltage control also removes the possibility of shock for persons operating the controls.

TROUBLESHOOTING

Problems in air conditioners can be caused by a number of malfunctioning parts. We have already mentioned how the thermostat, overload protector, start relay on larger units, and fan motor work. Knowing how they work means a lot in troubleshooting. If you know *how* they work, you can more easily identify *why* they are not working.

Throughout the industry you will find Problem–Probable Cause–Remedy charts made available by manufacturers of equipment. These are an aid to troubleshooting. They help in locating the possible trouble component or system. Then you need to identify the schematic and read it properly to be able to identify the problems associated with the electrical system.

Table 16-1 is a Problem-Probable Cause-Remedy listing of possible troubles with the hermetic compressors used in air-conditioning units. Take a close look at it and determine which are the troubles caused by electrical failure and those caused by the refrigerant system.

REVIEW QUESTIONS

1. What is meant by modes of cooling?
2. Where is the temperature sensor located on an air conditioner?
3. How do you make a high-speed fan run slower?
4. What is an auxiliary winding? What purpose does it serve?
5. What is the purpose of a run capacitor in the compressor circuit?
6. What is the purpose of a ladder diagram?
7. What is the function of the fan switch with AUTO and ON positions?
8. What is the purpose of the overload protector in series with the compressor motor?

PERFORMANCE OBJECTIVES

Know three different types of refrigerator thermostats.
Know how short relays work.
Know how defrost circuits are timed.
Know how various defrost timers operate.
Know how to use the ladder diagram to aid in trouble shooting a refrigerator.
Know how to locate defrost thermostats on refrigerators.
Know how hot gas defrost works.
Know how to troubleshoot using a Probable Cause-Remedy chart.

CHAPTER 17

Refrigeration Circuits

Circuits for refrigerators vary slightly according to the manufacturer. The best source for these schematics is the manufacturer's service manuals and bulletins. In this chapter we will be concerned primarily with the circuits of domestic (home) refrigerators with freezers as part of the unit.

BASIC REFRIGERATOR

The basic or simple refrigerator has an electrical circuit that consists of a plug, terminal board or plug-in connector, a start relay with overload protector, and a thermostat or temperature control. A light switch and lamp are usually included in the basic type of circuit (see Fig. 17-1).

Thermostats

Thermostats can have a number of design configurations, depending on the manufacturer. Figure 17-2 shows three different types of thermostats. Each has a knob on the end of the shaft when installed in a refrigerator.

Start Relay

The design of a start relay varies with the manufacturer of the device. Figures 17-3 and 17-4 show two types used for refrigerators. They are usually located on the compressor. S stands for start winding connection, C for the common line, and R for the run winding connection to the compressor motor.

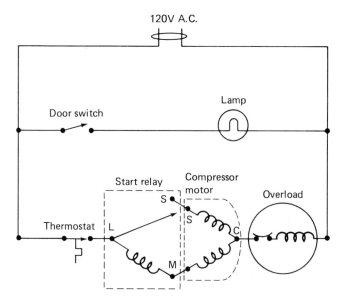

FIGURE 17-1 Basic refrigerator circuit.

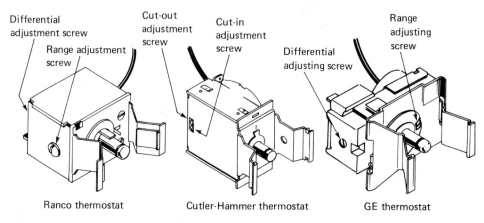

FIGURE 17-2 Three different types of thermostats. (Courtesy of Kelvinator)

REFRIGERATOR-FREEZER COMBINATION

Manual Defrost

The refrigerator-freezer combination exists in a manual defrost model for those interested in the most inexpensive device available. It has only the addition of a fan in the freezer compartment and may or may not have a light in the freezer section (see Fig. 17-5).

REFRIGERATOR–FREEZER COMBINATION

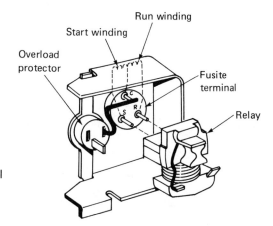

FIGURE 17-3 Fusite thermal relay and overload protector. (Courtesy of Kelvinator)

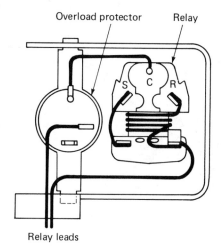

FIGURE 17-4 Start relay and overload protector. (Courtesy of Kelvinator)

Automatic Defrost

Frost-free refrigerators and freezers call for some means of timing the defrost cycle. As the frost accumulates in the freezer and refrigerator sections, it must be removed. It is easily removed if done often enough to avoid a buildup.

The defrost timer is used to control how often the defrost cycle takes place (see Fig. 17-6). Contacts 1 and 3 are used for the defrost motor coils. Contacts 2 and 4 are used to make contact with the proper circuit components to provide energy to the defrost heater or compressor (see Fig. 17-7).

The termination thermostat shown in the circuit is located near or on the evaporator to sense its temperature and indicate that the defrost cycle is complete and has done its job of removing the frost from the evaporator. The defrost *timer* can

326 REFRIGERATION CIRCUITS

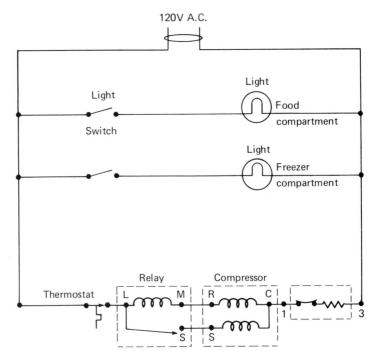

FIGURE 17-5 Basic refrigerator-freezer circuit.

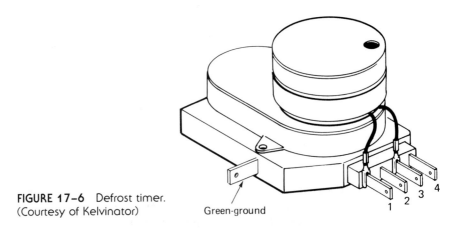

FIGURE 17-6 Defrost timer.
(Courtesy of Kelvinator)

be located in any number of places. Figure 17-8 illustrates how it may be placed on the back of the refrigerator.

A wiring diagram of a frost-free refrigerator is shown in Fig. 17-9. Note the addition of other features, such as a drip catcher heater, mullion heater, and door heater. These heaters serve different purposes. The drip catcher heater allows the

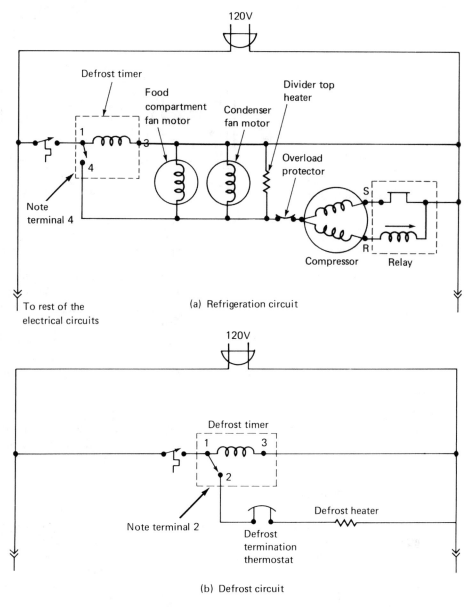

FIGURE 17-7 Note the refrigeration-cycle and the defrost-cycle circuits.

water that forms during defrost to drain out of the freezer unit to a pan underneath to be evaporated by the heat of the compressor nearby. The mullion heater and door heater are used to prevent a frost buildup on and around the doors of the freezer and refrigerator compartments when the outside air is very humid.

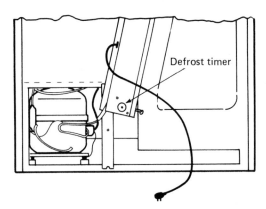

FIGURE 17-8 Location of the defrost timer on the refrigerator. (Courtesy of Kelvinator)

DEFROSTING

The timing of defrost operations is very important. Too long and you waste energy. Too short a time will result in a job half-done, with the resulting decrease in efficiency of the unit since the evaporator is loaded with frost and its ability to cool the compartment is diminished. A number of methods are utilized to defrost. They are all electrically controlled and therefore part of the circuits associated with refrigerator–freezers.

Defrost Thermostat

Defrost thermostats are precision built and tested to within ±6°F of the specified limit. A unique characteristic of a bimetal disc is that its calibration is fixed and does not change (see Fig. 17-10). This provides for reliability in excess of the life expectancy of the refrigerator or freezer. Continual life tests at the factory of 100,000 cycles, corresponding to 100 years, reveal that the calibration of a defrost thermostat does not drift out of tolerance. A slight creep of about 2°F occurs at about 28,000 cycles during the life of all defrost thermostats as the components wear in. A defrost thermostat can, however, be incorrectly calibrated from the beginning. But defrost thermostats never get weak during the life of the refrigerator or freezer. Therefore, if the defrost system has functioned properly for several months before failure, disregard the possibility of an incorrectly calibrated defrost thermostat. Do not suspect that the defrost thermostat has the wrong calibration unless *residual ice* is found in the evaporator.

The only method for checking the defrost thermostat in the field is to test it for continuity. The contacts should be closed at all times—except during the later part of the defrost cycle and for the first 10 minutes thereafter, when the compressor resumes operation.

You can determine that the defrost thermostat contacts are closed on side-by-side models by feeling the mullion for heat. On side-by-side models, the mullion

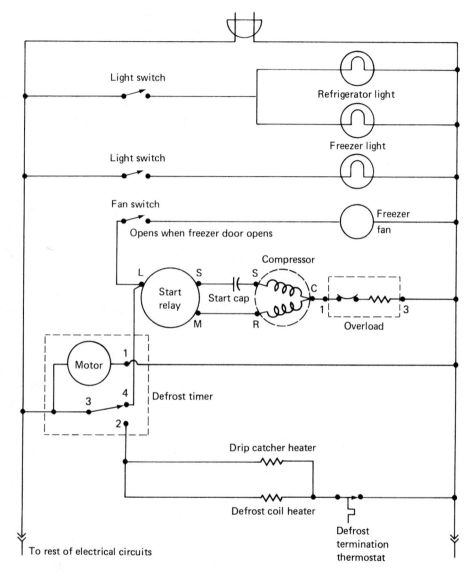

FIGURE 17-9 Refrigerator-freezer circuit.

heater is in series with the defrost thermostat. Thus, if the mullion heater is warm, the thermostat contacts must be closed (see Fig. 17-11).

Defrost Cycle

Resistance heat is used in automatic defrosting. A radiant heater is attached to the bottom of the evaporator with two aluminum straps. A defrost heater and defrost

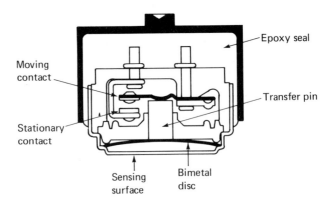

FIGURE 17-10 Cutaway view of a defrost termination thermostat. (Courtesy of General Electric)

termination thermostat are wired in series with the defrost heater. The defrost timer initiates and terminates a 21-minute defrost cycle every 6 hours. A termination thermostat cycles the defrost heater off at a predetermined temperature, prior to termination of the 21-minute defrost cycle. The aluminum drain positioned below the evaporator is defrosted by the defrost heater. Some refrigerators have a 25-minute defrost cycle. Another manufacturer makes models with a 17-minute cycle every 8 hours. As you can see, the timer is important in making sure the defrost takes place when needed to keep the evaporator free of frost.

Another method used is the *accumulated compressor time*. In this case, the defrost operates whether there is any frost on the evaporator or not. The amount of frost buildup is a function of the amount of time the compressor operates. Therefore, it is only reasonable that the accumulated time that the compressor operates is proportional to the frost buildup. This means of defrost timing is typical in some manufacturers' models.

The defrost timer motor in Fig. 17-12 is controlled by the refrigerator thermostat. The defrost timer motor operates only when the refrigerator thermostat calls for compressor operation. After about 6 hours of accumulated compressor operation, the defrost timer switch cam has rotated sufficiently to open the connection between terminals 1 and 4 of the timer. The compressor stops while the refrigerator thermostat continues to call for compressor operation. The circuit through the defrost timer now uses terminals 1 and 2. Power is fed to the defrost heater instead of the compressor. The switch remains in this position (causing the heater to warm up) for approximately 25 minutes. During this time period, the cam on the timer motor has been rotating. This rotation causes the switch to flip from terminal 2 to terminal 4. This action allows power to be restored from terminal 1 to terminal 4. The compressor can now operate and cool until the refrigerator thermostat switch opens.

One other method used to defrost is the *accumulated compressor time hot-gas*

DEFROSTING 331

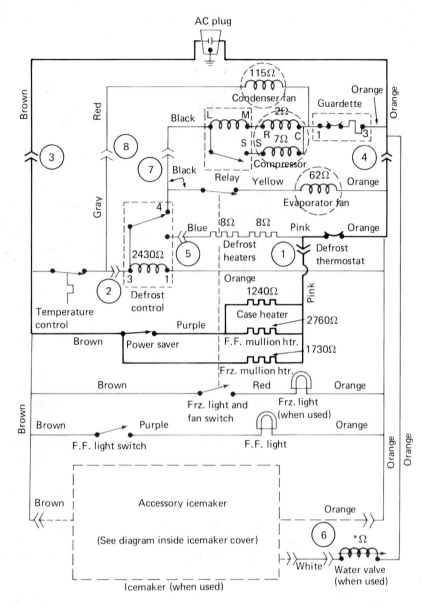

FIGURE 17-11 Schematic for the refrigerator–freezer. (Courtesy of General Electric)

332 REFRIGERATION CIRCUITS

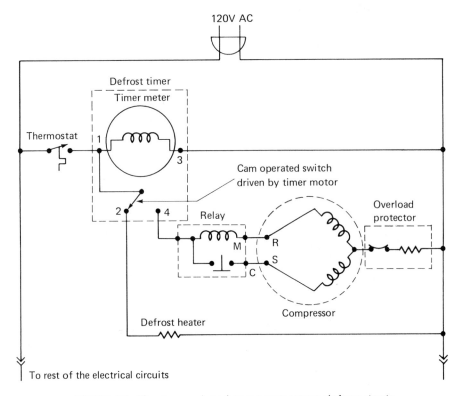

FIGURE 17-12 Accumulated compressor time defrost circuit.

defrost. This type of defrosting uses hot gas from the compressor to melt the frost accumulation on the evaporator. The defrost solenoid is important in this operation (see Fig. 17-13). The solenoid is used to open the line from the compressor to the piping alongside the evaporator, allowing hot gas to flow in the line. This piping is placed there originally by the manufacturer of the refrigerator. As the hot gas circulates through the piping, it causes the frost on the evaporator to melt. The compressor and evaporator are still in the refrigeration mode, and the cooling process continues while defrosting takes place. Timing is a function of the defrost timer motor and its cams.

OTHER DEVICES

As the refrigerator progressed from an "ice-box" replacement to one with a number of conveniences, other circuit changes were necessitated. In most instances they were add-ons. Some of the add-ons were the automatic ice maker, juice dispenser, soft

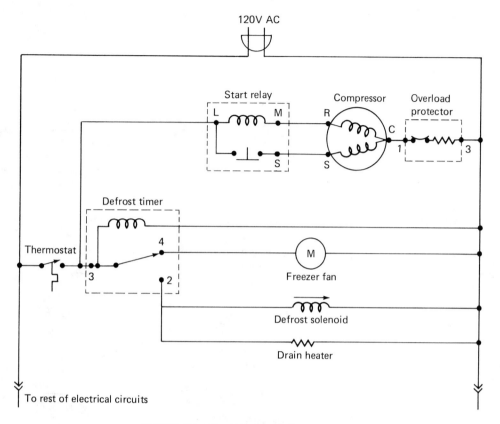

FIGURE 17-13 Hot-gas defrost circuit.

drink dispenser, and cold water faucets. Each refrigerator manufacturer has its own way of designing these devices. You should refer to the manufacturer's manuals and bulletins to be sure of the particular model's electrical circuits. See Fig. 17-11 for an illustration of where the ice maker is placed in the circuit.

Operation of the ice maker can be seen in Fig. 17-14. In Fig. 17-14A, note that near the completion of the first revolution the timing cam closes the water valve switch. However, since the thermostat is still closed, the mold heater circuit is energized. Current will not pass through the water valve solenoid and its switch since electrical current follows the path of least resistance.

At the end of the first revolution (Fig. 17-14B), the timing cam opens the holding switch. However, since the thermostat is still closed, a second revolution begins.

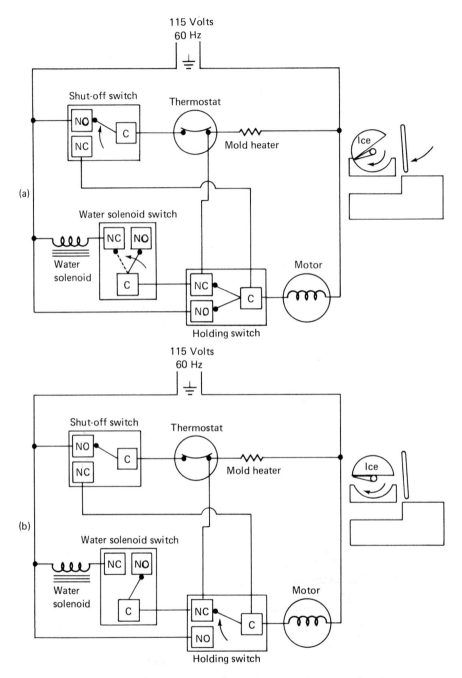

FIGURE 17-14 Circuit for ice maker.

TROUBLESHOOTING

To see the importance of the electrical circuits used in refrigerators and freezers, study Table 17-1. It shows some of the troubles that occur in a simple domestic freezer. *The electrical problems have been emphasized.* Compare these with what happens with the mechanical or refrigerant part of the freezer.

TABLE 17-1 Troubleshooting Freezers: Upright Models

Trouble	Probable Cause	Remedy or Repair
Product too cold.	1. Temperature selector knob set too cold.	1. Set warmer.
	2. Thermostat bulb contact bad.	2. If the bulb contact is bad, the bulb temperature will lag behind the cooling coil temperature. The unit will run longer and make the freezer too cold. See that the bulb makes good contact with the bulb well.
	3. Thermostat is out of adjustment.	3. Readjustment or change the thermostat.
Product too warm.	1. Thermostat selector knob set too warm.	1. Set cooler.
	2. Thermostat contact points dirty or burned.	2. Replace thermostat.
	3. Thermostat out of adjustment.	3. Readjust or change the thermostat.
	4. Loose electrical connection.	4. This may break the circuit periodically and cause the freezer to become warm because of irregular or erratic operation. Check the circuit and repair or replace parts.
	5. Excessive service load or abnormally high room temperature.	5. Unload part of the contents. Move unit to a room with lower temperature or exhaust excess room heat.
	6. Restricted air circulation over wrapped condenser.	6. Allow 6-inch clearance above the top and 3½-inch clearance at the sides and between the back of the cabinet and the wall.

(continued)

TABLE 17-1 (continued)

Trouble	Probable Cause	Remedy or Repair
	7. Excessive frost accumulation on the refrigerated shelves (manual defrost models).	7. Remove the frost.
	8. Compressor cycling on overload protector.	8. Check the protector and line voltage at the compressor.
Unit will not operate.	1. Service cord out of wall receptacle.	1. Plug in the service cord.
	2. Blown fuse in the feed circuit	2. Check the wall receptacle with a test lamp for a live circuit. If the receptacle is dead but the building has current, replace the fuse. Determine the cause of the overload or short circuit.
	3. Bad service cord plug, loose connection, or broken wire.	3. If the wall receptacle is live, check the circuit and make necessary repairs.
	4. Inoperative thermostat.	4. Power element may have lost charge or points may be dirty. Check the points. Short out the thermostat. Repair or replace the thermostat.
	5. Inoperative relay.	5. Replace.
	6. Stuck or burned-out compressor.	6. Replace the compressor.
	7. Low voltage. Cycling on overload.	7. Call utility company, asking them to increase voltage to the house. Or move unit to a separate household circuit.
	8. Inoperative overload protector.	8. Replace.
Unit runs all the time.	1. Thermostat out of adjustment.	1. Readjust or change the thermostat.
	2. Short refrigerant charge (up to 4 oz). Cabinet temperatures abnormally low in lower section.	2. Not enough refrigerant to flood the evaporator coil at the outlet to cause the thermostat to cut out. Recharge and test for leaks.
	3. Restricted air flow over the wrapper condenser.	3. Provide proper clearances around the cabinet.
	4. Inefficient compressor.	4. Replace.

TABLE 17-1
(continued)

Trouble	Probable Cause	Remedy or Repair
Unit short cycles.	1. Thermostat erratic or out of adjustment.	1. Readjust or change the thermostat.
	2. Cycling on the relay.	2. This may be caused by low or high line voltage that varies more than 10% from the 115 volts. It may also be caused by high discharge pressures caused by air or noncondensable gases in the system. Correct either condition.
Unit runs too much.	1. Abnormal use of the cabinet.	1. Heavy usage requires more operation. Check the usage and correct or explain.
	2. Shortage of refrigerant.	2. Unit will run longer to remove the necessary amount of heat and it will operate at a lower than normal suction pressure. Put in the normal charge and check for leaks.
	3. Overcharge of refrigerant.	3. Excessively cold or frosted suction line results in lost refrigeration effort. Unit must run longer to compensate for the loss. Purge off excessive charge.
	4. Restricted air flow over the condenser.	4. This can result if the cabinet is enclosed. This will obstruct the air flow around the cabinet shell. Restricted air flow can also be caused by air or noncondensable gases in the system. This results in a higher head pressure. The higher head pressure produces more re-expansion during the suction stroke of the compressor. Consequently, less suction vapor is taken. Increased

(continued)

338 REFRIGERATION CIRCUITS

TABLE 17-1 (continued)	Trouble	Probable Cause	Remedy or Repair
			running time must compensate for loss of efficiency. Correct the condition.
		5. High room or ambient temperature.	5. Any increase in temperature around the cabinet will increase the refrigeration load. This will result in longer running time to maintain cabinet temperature.
	Too much frost on refrigerated surfaces—low-side.	1. Abnormally heavy usage in humid weather.	1. Do not leave the freezer door open any longer than necessary to load or remove products.
		2. Poor door gasket seals.	2. This permits the entrance of moisture by migration, which freezes out of the air as frost on the refrigerated surfaces.

Courtesy of Kelvinator.

RAPID ELECTRICAL DIAGNOSIS

General Electric and Hotpoint have a quick and accurate device for diagnosing electrical faults in a refrigerator. The hand-held device is a result of down-sizing a computer used in the factory to check production models of refrigerators. The device is called "Big Red" for Rapid Electrical Diagnosis unit (see Fig. 17-15). It allows the technician to check almost every electrical component in the refrigerator within 6 minutes. It is not necessary to unplug the refrigerator, unload any food, move the refrigerator from the wall, or, usually, open the door.

Multiconnectors are designed into the refrigerator wiring harness. They are located behind the front grill. A mini-manual contains information that makes the device useful for a particular refrigerator model. All the latest models with the RED feature have a packet or envelope containing the necessary information. The packet contains the pictoral and schematic wirings, RED component circuits, machine wiring diagram, and components energized through the RED system (see Figs. 17-16 and 17-17).

Figure 17-16 shows the pictorial of the wiring harness and color code for the RED connector. Figure 17-17 gives the electrical schematic and beside it is the same

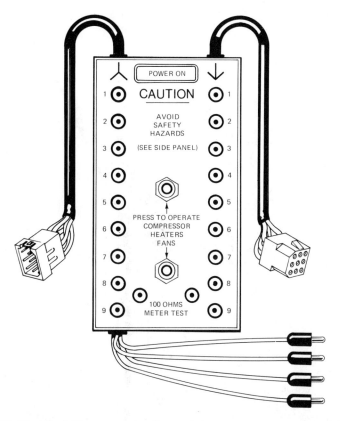

FIGURE 17-15 RED test unit for General Electric and Hotpoint refrigerators.

diagram, but with the $\prec$ and $\succ$ connections indicating whether it is a female or male type connection for that point.

Use of RED may reduce the technician's service time by 20%. It reduces the additional service callbacks for the same problem by 50%. All no-frost GE models 16 cubic feet and larger have this feature. Hotpoint uses the same system.

ENERGY-SAVER SWITCH

An energy-saver switch is the latest development in refrigerator design (see Fig. 17-18). The energy-saver switch can reduce operating costs by taking some of the heaters out of the circuit. Controls for the case and mullion antisweat heaters are incorporated into the energy-saver switch. The heaters prevent moisture formation on the outside of the case in humid conditions. The normal setting may be used about 80% of the time due to low humidity in the winter and air conditioning in

340 REFRIGERATION CIRCUITS

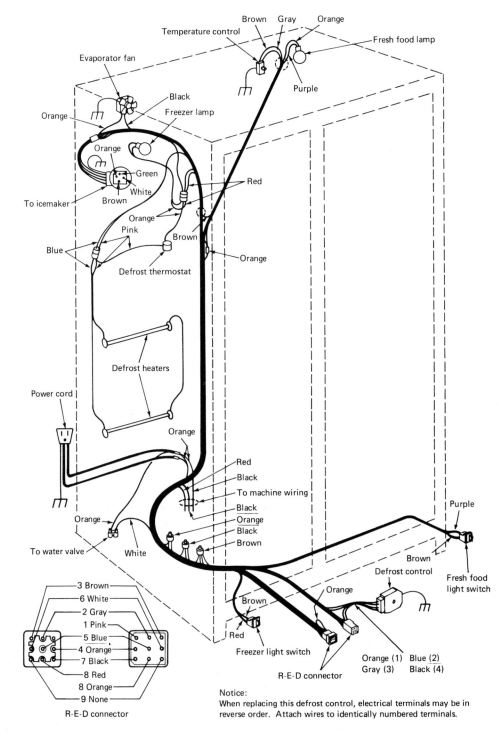

FIGURE 17-16 Pictorial drawing of the electrical wiring of a side-by-side GE 20-cubic-foot refrigerator that uses the RED system for troubleshooting.

ENERGY-SAVER SWITCH 341

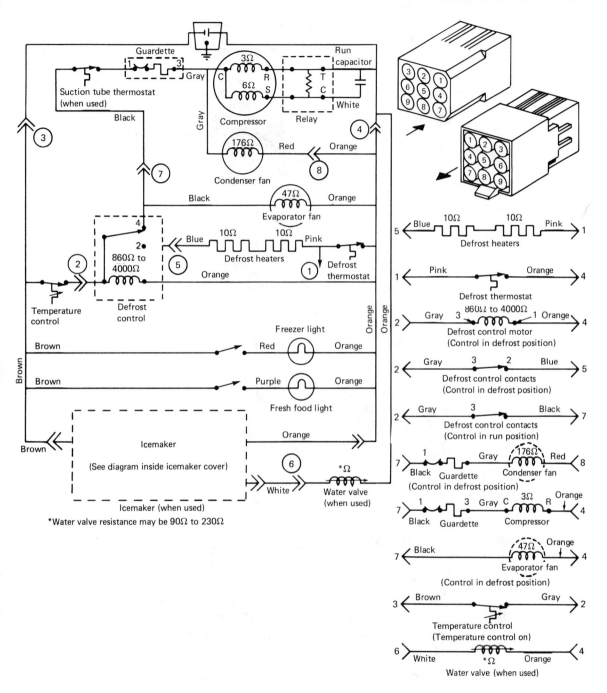

FIGURE 17-17 Ladder diagram of a GE 20-cubic-foot refrigerator that uses the RED system.

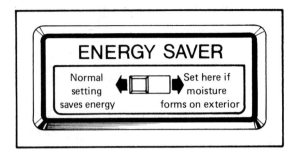

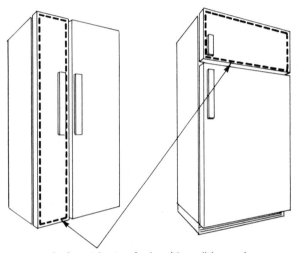

FIGURE 17-18 Energy-saver switches save electricity and are used in conjunction with antisweat heaters located around the freezer section doors.

Antisweat heaters for humid conditions only

the summer. The switch can also be set to an alternate setting only if moisture forms on the outside of the refrigerator.

REVIEW QUESTIONS

1. What makes up the basic refrigerator?
2. What is the purpose of the thermostat in a basic refrigerator?
3. What is the function of the start relay in a refrigerator?
4. What is necessary to make a refrigerator frost free?
5. How does the defrost timer work?
6. Why is a defrost thermostat needed in a refrigerator?
7. What is meant by accumulated compressor time hot-gas defrost?
8. How does the ice maker work in a refrigerator?

PERFORMANCE OBJECTIVES

Know some of the common sense rules for safely working around air-conditioning and refrigeration equipment.

Know how often filters on air conditioners have to be cleaned.

Know how the low supply voltage affects the operation of electrical equipment.

Know the source of most trouble in air-conditioning systems.

Know the source of most trouble in heating systems.

Know how to test motor capacitors.

Know how to use a clamp-on meter to troubleshoot.

Know how to test the centrifugal switch in a single-phase motor.

Know how to measure the capacitance of a capacitor in a piece of heating and cooling equipment.

Know good troubleshooting procedures.

CHAPTER 18

Troubleshooting

From time to time, air-conditioning and refrigeration units experience problems with the electrical and the mechanical aspects of their operation. They have in some cases been moved and damaged, and in other cases they have been repaired with components that are incorrect or wired into the circuits incorrectly.

The service technician must be aware of the problems that may develop in a period of time so that they may be located quickly, easily, and without undue expense to the owner of the unit. In this chapter we will attempt to show how troubleshooting procedures can aid in making the job faster and easier.

SAFETY

Safety first has its direct and implied meaning. You can work with air-conditioning and refrigeration equipment safely if a few commonsense rules are followed.

Handling Refrigerants

The proper use of gloves, eye protection, and clothing to protect the body is necessary. Freon escaping from a refrigeration unit can cause permanent damage to the skin. It can also cause blindness if you are hit in the face with the gases under pressure. Make sure you wear the proper clothing at all times when troubleshooting or recharging a unit.

Testing Precaution

Pressure testing or cleaning refrigeration and air-conditioning systems can be dangerous. Be careful not to exceed 150 psig when pressure testing a complete system.

Electrical Safety

The main rule for electrical safety is to see to it that the main circuit breaker is in the off position and locked before starting to remove or check (with an ohmmeter) any refrigeration, air-conditioning, or heating equipment. Keep in mind that some compressors have power applied at all times to the off-cycle crankcase heater. Even if the compressor is not running, the power is applied to the crankcase heater. Some run capacitors are connected to the compressor motor windings even when the compressor is not running. Other devices are energized when the compressor is not running. Thus, electrical power is applied to the unit even when the compressor is not running.

COMPRESSOR PROBLEMS

There are a number of compressor problems that can be quickly identified from a table of problems, possible causes, and suggested repair (see Table 18-1). The compressor is the heart of the refrigeration system whether it is the air conditioner, refrigerator, or freezer. That makes it of primary concern in any troubleshooting procedure.

TABLE 18-1 Compressor Troubleshooting and Service

Complaint	Possible Cause	Repair
Compressor will not start. There is no hum.	1. Line disconnect switch open.	1. Close start or disconnect switch.
	2. Fuse removed or blown.	2. Replace fuse.
	3. Overload protector tripped.	3. Refer to electrical section.
	4. Control stuck in open position.	4. Repair or replace control.
	5. Control off due to cold location.	5. Relocate control.
	6. Wiring improper or loose.	6. Check wiring against diagram.
Compressor will not start. It hums, but trips on overload protector.	1. Improperly wired.	1. Check wiring against diagram.
	2. Low voltage to unit.	2. Determine reason and correct.
	3. Starting capacitor defective.	3. Determine reason and replace.
	4. Relay failing to close.	4. Determine reason and correct, replace if necessary.

TABLE 18-1 (continued)

Complaint	Possible Cause	Repair
	5. Compressor motor has a winding open or shorted.	5. Replace compressor.
	6. Internal mechanical trouble in compressor.	6. Replace compressor.
	7. Liquid refrigerant in compressor.	7. Add crankcase heater and/or accumulator.
Compressor starts, but does not switch off of start winding.	1. Improperly wired.	1. Check wiring against diagram.
	2. Low voltage to unit.	2. Determine reason and correct.
	3. Relay failing to open.	3. Determine reason and replace if necessary.
	4. Run capacitor defective.	4. Determine reason and replace.
	5. Excessively high discharge pressure.	5. Check discharge shut-off valve, possible overcharge, or insufficient cooling of condenser.
	6. Compressor motor has a winding open or shorted.	6. Replace compressor.
	7. Internal mechanical trouble in compressor (tight).	7. Replace compressor.
Compressor starts and runs, but short cycles on overload protector.	1. Additional current passing through the overload protector.	1. Check wiring against diagram. Check added fan motors, pumps, etc., connected to wrong side of protector.
	2. Low voltage to unit (or unbalanced if three phase).	2. Determine reason and correct.
	3. Overload protector defective.	3. Check current, replace protector.
	4. Run capacitor defective.	4. Determine reason and replace.
	5. Excessive discharge pressure.	5. Check ventilation, restrictions in cooling medium, restrictions in refrigeration system.
	6. Suction pressure too high.	6. Check for possibility of misapplication. Use stronger unit.
	7. Compressor too hot—return gas hot.	7. Check refrigerant charge. (Repair leak.) Add refrigerant if necessary.
	8. Compressor motor has a winding shorted.	8. Replace compressor.

(continued)

TABLE 18-1 (continued)

Complaint	Possible Cause	Repair
Unit runs, but short cycles on.	1. Overload protector	1. Check current. Replace protector.
	2. Thermostat	2. Differential set too close. Widen.
	3. High-pressure cutout due to insufficient air or water supply, overcharge, or air in system.	3. Check air or water supply to condenser. Reduce refrigerant charge, or purge.
	4. Low-pressure cutout due to:	4.
	a. Liquid line solenoid leaking.	a. Replace.
	b. Compressor valve leak.	b. Replace.
	c. Undercharge.	c. Repair leak and add refrigerant.
	d. Restriction in expansion device.	d. Replace expansion device.
Unit operates long or continuously.	1. Shortage of refrigerant.	1. Repair leak. Add charge.
	2. Control contacts stuck or frozen closed.	2. Clean contacts or replace control.
	3. Refrigerated or air-conditioned space has excessive load or poor insulation.	3. Determine fault and correct.
	4. System inadequate to handle load.	4. Replace with larger system.
	5. Evaporator coil iced.	5. Defrost.
	6. Restriction in refrigeration system.	6. Determine location and remove.
	7. Dirty condenser.	7. Clean condenser.
	8. Filter dirty.	8. Clean or replace.
Start capacitor open, shorted, or blown.	1. Relay contacts not operating properly.	1. Clean contacts or replace relay if necessary.
	2. Prolonged operation on start cycle due to:	2.
	a. Low voltage to unit.	a. Determine reason and correct.
	b. Improper relay.	b. Replace.
	c. Starting load too high.	c. Correct by using pump-down arrangement if necessary.
	3. Excessive short cycling	3. Determine reason for short cycling as mentioned in previous complaint.
	4. Improper capacitor.	4. Determine correct size and replace.

TABLE 18-1 (continued)

Complaint	Possible Cause	Repair
Run capacitor open, shorted, or blown.	1. Improper capacitor. 2. Excessively high line voltage (110% of rated maximum).	1. Determine correct size and replace. 2. Determine reason and correct.
Relay defective or burned out.	1. Incorrect relay. 2. Incorrect mounting angle. 3. Line voltage too high or too low. 4. Excessive short cycling. 5. Relay being influenced by loose vibrating mounting. 6. Incorrect run capacitor.	1. Check and replace. 2. Remount relay in correct position. 3. Determine reason and correct. 4. Determine reason and correct. 5. Remount rigidly. 6. Replace with proper capacitor.
Space temperature too high.	1. Control setting too high. 2. Expansion valve too small. 3. Cooling coils too small. 4. Inadequate air circulation.	1. Reset control. 2. Use larger valve. 3. Add surface or replace. 4. Improve air movement.
Suction line frosted or sweating.	1. Expansion valve oversized or passing excess refrigerant. 2. Expansion valve stuck open. 3. Evaporator fan not running. 4. Overcharge of refrigerant.	1. Readjust valve or replace with smaller valve. 2. Clean valve of foreign particles. Replace if necessary. 3. Determine reason and correct. 4. Correct charge.
Liquid line frosted or sweating.	1. Restriction in dehydrator or strainer. 2. Liquid shutoff (king valve) partially closed.	1. Replace part. 2. Open valve fully.
Unit noisy.	1. Loose parts or mountings. 2. Tubing rattle. 3. Bent fan blade causing vibration. 4. Fan motor bearings worn.	1. Tighten. 2. Reform to be free of contact. 3. Replace blade. 4. Replace motor.

Courtesy of Kelvinator.

PSC Compressors

The permanent split-capacitor compressor has some problems that should be uppermost in your mind so that you are aware of them (see Table 18-2). The low-voltage problems are listed and possible corrections given. The branch circuit fuse or circuit breaker trips and causes the unit to become disabled. The possible cause could be that the rating of the protection device is not high enough to handle the current. Some problems are within the scope of the work the technician can handle. Others, like low line voltage caused by factors beyond the control of the homeowner, call for the utility company to correct them.

A lot of troubleshooting is common sense. If the fuse is blown and you put in another and it blows, it tells you that there is something drawing too much current. Simply check for cause.

TABLE 18-2 PSC Compressor Motor Troubles and Corrections

Causes	Corrections
Low Voltage	
1. Inadequate wire size.	1. Increase wire size.
2. Watt-hour meter too small.	2. Call utility company.
3. Power transformer too small or feeding too many homes.	3. Call utility company.
4. Input voltage too low.	4. Call utility company.
(Note: Starting torque varies as the square of the input voltage.)	
Branch Circuit Fuse or Circuit Breaker Tripping	
1. Rating too low.	1. Increase size to a minimum of 175% of unit FLA (full load amperes) to a maximum of 225% of FLA.
System Pressure High or Not Equalized	
1. Pressures not equalizing within three minutes.	1. a. Check metering device (capillary tube or expansion valve). b. Check room thermostat for cycling rate. Off cycle should be at least 5 minutes. Also check for "chattering." c. Has some refrigerant dryer or some other possible restriction been added?
2. System pressure too high.	2. Make sure refrigerant charge is correct.
3. Excessive liquid in crankcase (split-system applications).	3. Add crankcase heater and suction line accumulator.
Miscellaneous	
1. Run capacitor open or shorted.	1. Replace with new, properly sized capacitor.
2. Internal overload open.	2. Allow 2 hours to reset before changing compressor.

Air-conditioner Compressors

Home air conditioners, those that fit into a window or those that are part of a central air-conditioning system, are subject to problems with their compressors and electrical controls. Most of those that operate in the window are designed for 120- or 240-volt operation. In some cases, during the summer when line voltage is low, they may experience some low-voltage problems.

The units are equipped with some kind of filter to make sure the air is cleaned before it is forced into the room being cooled. These filters usually require a change or cleaning at least once a year. Where dust is a problem, such maintenance should be more frequent. At this time, the condenser coil should be brushed with a soft brush and flushed with water. The filters should be vacuumed and then washed to remove dust. The outside of the case should be wiped clean with a soapy cloth. The cleaner the filter, the more efficient the unit. In some cases a clogged filter can cause compressor problems. Table 16-1 shows some of the problems associated with the compressor in these units. Note how most of the problems associated with air conditioners are electrical.

LOW-VOLTAGE OPERATION

Electrical apparatus designed to produce at full capacity at the voltage indicated on the rating plate can malfunction at lower than designated voltages. Motors operated at lower than rated voltage cannot provide full horsepower without shortening their service life. Relays and solenoids can also fail to operate if low voltage is present.

The Air Conditioning and Refrigeration Institute (ARI) certifies the cooling units after testing them. The units are tested to make sure they will operate with 10% above or below rated voltage. This does not mean they will operate continuously without damage to the motor. Most air-conditioning compressor burnouts are caused by low-voltage operation. A hermetic compressor is entirely enclosed within the refrigerant cycle; it is very important that it not be abused by overloading or low voltage. Both conditions can occur during peak load periods. A national survey has shown that the most common cause of compressor low voltage is the use of undersized conductors between the utility lines and the condensing unit.

USING A SYSTEM TO TROUBLESHOOT (ELECTRICAL)

Refrigeration, air-conditioning, and heating units all have various systems that cause them to operate and do the job they were designed for. First, you have to identify the system and the problems most commonly associated with the particular system. Then you have to be able to determine exactly what caused the symptom you are witnessing. Next, you have to be able to correct the problem and put the unit back into operation.

352 TROUBLESHOOTING

Electrical tests are the most common because the electrical problems are the most frequently encountered in all three types of equipment. A systematic procedure is necessary to obtain the needed results.

Motor Testing

Testing a motor in a sealed condensing unit to determine why it does not operate becomes a very simple process if the correct procedure is followed. Each test made should be one of a series of eliminations to determine what part of the system is defective. By checking other parts of the wiring system before checking the unit itself, a great deal of time can be saved since, in most cases, the trouble will be in the wiring or controls rather than in the unit.

To make a complete electrical test on electrical outlets and on the unit itself, it is advisable to make a test cord (see Fig. 18-1). By connecting the black and white terminals together and placing a light bulb in the socket, the cord may be used to check the wall outlet into which the unit is connected. By connecting the white and red terminal clips, this same test may be made by depressing the push button. This will serve as a test to make certain the push button is in working order. This test cord should have a capacitor installed in the red lead to the push button if the compressor is a capacitor-start type.

When these tests have been completed and it is known that current is being supplied to the unit, the next step is to check the three wires on the base of the compressor unit. Pull the plug from the wall receptacle and carefully examine the nuts that hold the wires in position. Then try each wire to be sure it is held firmly in place, as a loose wire may keep the unit from operating. Test the thermostatic switch to determine whether or not contact is being made at that point. Turn the cold-control knob several times. If this fails to start the unit, then short across the thermostatic switch terminals on the switch. To do this, it will be necessary to remove the switch cover from the top of the switch. If the unit starts, it is an indication that the thermostatic switch is not operating properly and must be repaired or replaced. After the thermostatic switch has been checked and if the trouble is not

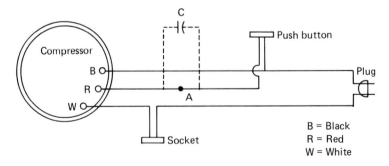

FIGURE 18-1 Test lamp for troubleshooting.

located, it will be necessary to determine whether the trouble is in the motor, motor-protector relay, or capacitor.

Capacitor Testing

The capacitor must be checked before testing the unit itself. This is done in the following manner:

1. Disconnect the capacitor wires from the motor-protector relay.
2. Connect these two wires to the black and white terminals of the test cord.
3. Put a 150-watt light bulb in the receptacle on the test cord, and plug it into an outlet.

If the 150-watt bulb does not light, it is an indication that the capacitor has an open circuit and must be replaced. However, if the bulb does light, it is not an indication that the capacitor is perfect. This must be checked further by shorting across the two terminals of the capacitor with a screwdriver with an insulated handle. If the brilliance of the light changes, that is, if the light bulb burns brighter when the terminals of the capacitor are shorted, it indicates that the capacitor is in proper operating condition. A decided sparking of the terminals will also be noticed when the terminals are shorted. If the brilliance of the bulb does not change, it is an indication that the capacitor has an internal short and must be replaced.

Motor-protector Relay Testing

If the overload protector is found defective during the preceding test, replace it (see Fig. 18-2). The motor overload protector is usually accessible for replacement and

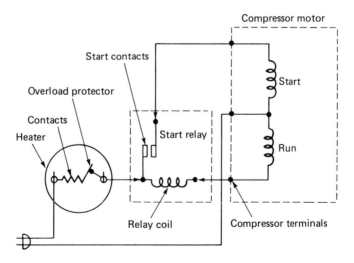

FIGURE 18-2 Location of overload protector in the circuit.

354 TROUBLESHOOTING

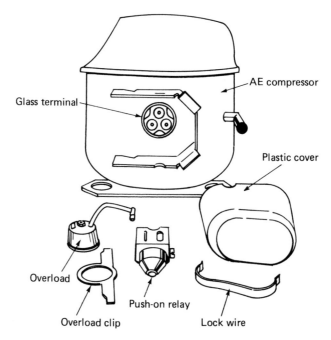

FIGURE 18-3 Compressor with its glass terminal and associated components. (Courtesy of Tecumseh)

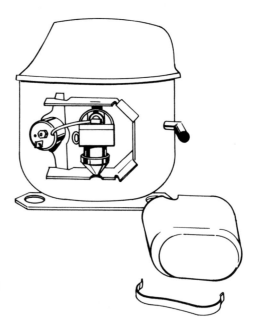

FIGURE 18-4 Overload protector location on the compressor. (Courtesy of Tecumseh)

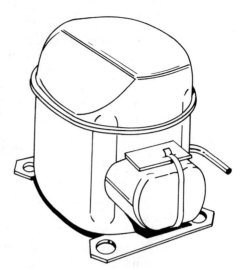

FIGURE 18-5 Compressor with cover. (Courtesy of Tecumseh)

is located near the compressor (see Fig. 18-3). Remove the clips holding the overload in position, and remove the wire from the terminal by pulling it outward. Notice the position of the electrical lead before replacing it (see Fig. 18-4). Place the electrical leads on the replacement overload and check for correct connections against the wiring diagram. Close up the unit (see Fig. 18-5). Test for proper operation.

USING METERS TO CHECK FOR PROBLEMS

The voltmeter and the ohmmeter can be used to isolate various problems. You should be able to read the schematic and make the proper voltage or resistance measurements. An incorrect reading will indicate the possibility of a problem. Troubleshooting charts will aid in isolating the problem to a given system. Once you have arrived at the proper system that may be causing the symptoms noticed, you will then need to use the ohmmeter with the power off to isolate a section of the system. Once you have zeroed in on the problem, you can locate it by knowing what the proper reading should be. Deviation from a stated reading of over 10% is usually indicative of a malfunction, and in most cases the component part must be replaced to assure proper operation and no call backs.

USING A VOLT-AMMETER FOR TROUBLESHOOTING ELECTRIC MOTORS

Most electrical equipment will work satisfactorily if the line voltage differs ±10% from the actual nameplate rating. In a few cases, however, a 10% voltage drop may result in a breakdown. Such may be the case with an induction motor that is being loaded to its fullest capacity both on start and run. A 10% loss in the line voltage will result in a 20% loss in torque.

The full-load current rating on the nameplate is an approximate value based on the average unit coming off the manufacturer's production line. The actual current for any one unit may vary as much as ±10% at rated output. However, a load current that exceeds the rated value by 20% or more will reduce the life of the motor due to higher operating temperatures, and the reason for excessive current should be determined. In many cases it may simply be an overloaded motor. The percentage of increase in load will not correspond with percentage of increase in load current. For example, in the case of a single-phase induction motor, a 35% increase in current may correspond to an 80% increase in torque output.

The operating conditions and behavior of electrical equipment can be analyzed only by actual measurement. A comparison of the measured terminal voltage and current will check whether the equipment is operating within electrical specifications.

A voltmeter and an ammeter are needed for the two basic measurements. To measure voltage, the test leads of the voltmeter are in contact with the terminals of the line under test. To measure current, the conventional ammeter must be connected in series with the line so that the current will flow through the ammeter.

To insert the ammeter, you must shut down the equipment, break open the line, connect the ammeter, and then start up the equipment to read the meter. And you have to do the same to remove the meter once it has been used. Other time-consuming tests may have to be made to locate the problem. However, all this can be eliminated by the use of a clamp-on volt-ammeter.

Clamp-on Volt-Ammeter

The pocket size volt-ammeter shown in Fig. 18-6 is the answer to most troubleshooting problems on the job. The line does not have to be disconnected to obtain a current reading. The meter works on the transformer principle; it picks up the magnetic lines surrounding a current-carrying conductor and presents this as a function of the entire amount flowing through the line. Remember, in transformers we discussed how the magnetic field strength in the core of the transformer determines the amount of current in the secondary. The same principle is used here to detect the flow of current and how much.

To get transformer action, the line to be tested is encircled with the split-type core by simply pressing the trigger button. Aside from measuring terminal voltages and load currents, the split-core ammeter-voltmeter can be used to track down electrical difficulties in electric motor repair.

Looking for Grounds

To determine whether a winding is grounded or has a very low value of insulation resistance, connect the unit and test leads as shown in Fig. 18-7. Assuming the available line voltage is approximately 120 volts, use the unit's lowest voltage range. If the winding is grounded to the frame, the test will indicate full line voltage.

A high-resistance ground is simply a case of low insulation resistance. The

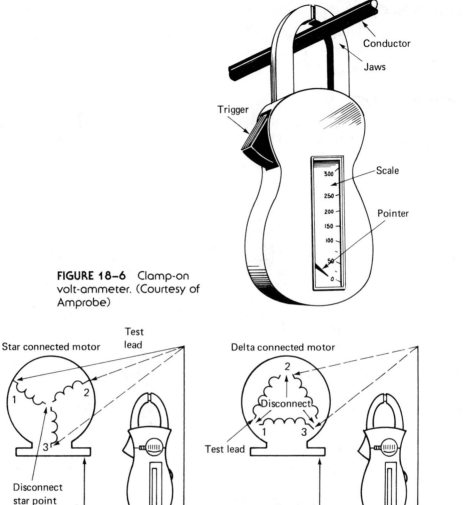

FIGURE 18-6 Clamp-on volt-ammeter. (Courtesy of Amprobe)

FIGURE 18-7 Grounded phase of a motor. (Courtesy of Amprobe)

indicated reading for a high-resistance ground will be a little less than line voltage. A winding that is not grounded will be evidenced by a small or negligible reading. This is due mainly to the capacitive effect between the windings and the steel lamination.

To locate the grounded portion of the windings, disconnect the necessary connection jumpers and test. Grounded sections will be detected by a full line voltage indication.

Looking for Opens

To determine whether a winding is open, connect test leads as shown in Figs. 18-8 and 18-9. If the winding is open, there will be no voltage indication. If the circuit is not open, the voltmeter indication will read full line voltage.

Looking for Shorts

Shorted turns in the winding of a motor behave like a shorted secondary of a transformer. A motor with a shorted winding will draw excessive current while running at no load. Measurement of the current can be made without disconnecting lines.

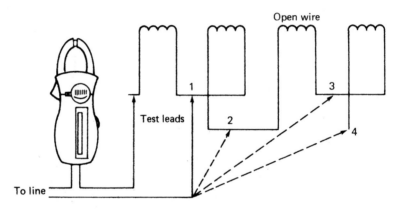

FIGURE 18-8 Isolating an open phase. (Courtesy of Amprobe)

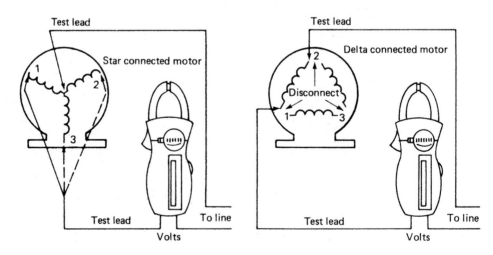

FIGURE 18-9 Finding an open phase. (Courtesy of Amprobe)

You engage one of the lines with the split-core transformer of the tester. If the ammeter reading is much higher than the full-load ampere rating on the nameplate, the motor is probably shorted.

In a two- or three-phase motor, a partially shorted winding produces a higher current reading in the shorted phase. This becomes evident when the current in each phase is measured.

MOTORS WITH SQUIRREL-CAGE ROTORS

Loss in output torque at rated speed in an induction motor may be due to opens in the squirrel-cage rotor. To test the rotor and determine which rotor bars are loose or open, place the rotor in a growler. Engage the split-core ammeter around the lines going to the growler, as shown in Fig. 18-10. Set the switch to the highest current range. Switch on the growler and then set the test unit to the approximate current range. Rotate the rotor in the growler and take note of the current indication whenever the growler is energized. The bars and end rings in the rotor behave similarly to a shorted secondary of a transformer. The growler windings act as the primary. A good rotor will produce approximately the same current indications for all positions of the rotor. A defective rotor will exhibit a drop in the current reading when the open bars move into the growler field.

TESTING THE CENTRIFUGAL SWITCH IN A SINGLE-PHASE MOTOR

A defective centrifugal switch may not disconnect the start winding at the proper time. To determine conclusively that the start winding remains in the circuit, place the split-core ammeter around one of the start-winding leads. Set the instrument to the highest current range. Turn on the motor switch. Select the appropriate current range. Observe if there is any current in the start-winding circuit. A current indication signifies that the centrifugal switch did not open when the motor came up to speed (see Fig. 18-11).

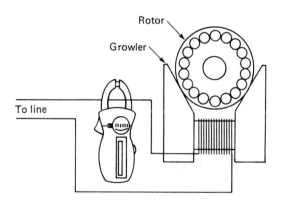

FIGURE 18-10 Using a growler to test a motor. (Courtesy of Amprobe)

360 TROUBLESHOOTING

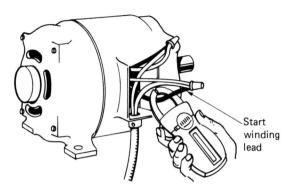

FIGURE 18-11 Checking the centrifugal switch with a clamp-on meter. (Courtesy of Amprobe)

TESTING FOR A SHORT CIRCUIT BETWEEN RUN AND START WINDINGS

A short between run and start windings may be determined by using the ammeter and line voltage to check for continuity between the two separate circuits. Disconnect the run and start winding leads and connect the instrument as shown in Fig. 18-12. Set the meter on voltage. A full-line voltage reading will be obtained if the windings are shorted to one another.

CAPACITOR TESTING

Defective capacitors are very often the cause of trouble in capacitor-type motors. Shorts, opens, grounds, and insufficient capacity in microfarads are conditions for which capacitors should be tested to determine whether they are good.

You can determine a grounded capacitor by setting the instrument on the proper voltage range and connecting it and the capacitor to the line as shown in Fig. 18-13. A full-line voltage indication on the meter signifies that the capacitor is

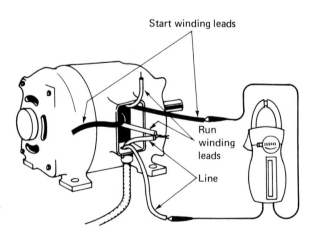

FIGURE 18-12 Finding a shorted winding using a clamp-on meter. (Courtesy of Amprobe)

CAPACITOR TESTING

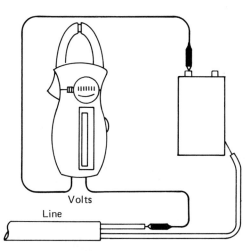

FIGURE 18-13 Finding a grounded capacitor with a clamp-on meter. (Courtesy of Amprobe)

grounded to the can. A high-resistance ground is evident from a voltage reading that is somewhat below the line voltage. A negligible reading or a reading of no voltage indicates that the capacitor is not grounded.

Measuring the Capacity of a Capacitor

To measure the capacity of the capacitor, set the test unit's switch to the proper voltage range and read the line voltage indication. Then set to the appropriate current range and read the capacitor current indication. During the test, keep the capacitor on the line for a very short period of time, because motor starting electrolytic capacitors are rated for intermittent duty (see Fig. 18-14). The capacity in microfarads is then computed by substituting the voltage and current readings in the following formula, assuming that a full 60-hertz line is used:

$$\text{microfarads} = \frac{2650 \times \text{amperes}}{\text{volts}}$$

An open capacitor will be evident if there is no current indication in the test. A shorted capacitor is easily detected. It will blow the fuse when the line switch is turned on to measure the line voltage. Troubleshooting can be broken down into a simple procedure. Logical thinking is what it takes to be able to accurately diagnose a problem and then correct it. For instance, you can isolate a particular system since you know how the system is supposed to operate. If it is not operating according to expectations, then you know something is causing the problem. By concentrating on a particular system that has been located, it is possible to concentrate your thinking on things that would cause the malfunction.

Once you have determined which components in the system can cause the symptoms being experienced, it is possible to find the faulty component.

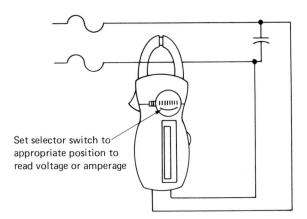

FIGURE 18-14 Finding the size of a capacitor with a clamp-on meter. (Courtesy of Amprobe)

After the problem has been located and the component identified, it is then necessary to take the component out and repair it or replace it.

Once the problem has been located, the part removed and a new one inserted or the old one repaired it is necessary to check the operation of the system to make sure it is performing according to its design characteristics.

All this can be broken down into key words that will aid you in any troubleshooting situation:

TROUBLESHOOTING PROCEDURE

Isolate. Isolate the system giving the trouble. The *system* may be:

the evaporator

the condenser

the compressor

the controls

Use the Problem-Probable Cause-Remedy charts to aid in isolating the problem.

Concentrate. Concentrate on the isolated system. Concentrate on *how* it works and *why* it would malfunction.

Differentiate. Determine which components could cause the identified symptoms; then locate each possible troublemaking part.

Eliminate. Eliminate the possible troublemakers one by one. Use the proper test instrument to do so.

Repair or Replace. Either repair or replace the identified troublemaker.

Check. Check to see if the unit operates properly with the repaired or replaced parts.

REVIEW QUESTIONS

1. What are some commonsense rules for safely working around air-conditioning and refrigeration equipment?
2. How often should filters on air conditioners be cleaned?
3. How does low supply voltage affect the operation of electrical equipment?
4. What does ARI mean? What is its function?
5. What is the source of most trouble in air-conditioning systems?
6. What is the source of most trouble in heating systems?
7. How do you test motor capacitors?
8. How do you use a clamp-on ammeter to test electric motors?
9. How do you test the centrifugal switch in a single-phase motor?
10. How do you measure the capacitance of a capacitor in a piece of heating and cooling equipment?
11. What is a good troubleshooting procedure?

Index

A

AC, 116, 135
  connecting, 116
  maximum and peak values, 138
  phase, 138
  polyphase, 139
    three-phase, 140
    connections, 141
  root-means-square (rms), 139
Actuators, 210
Air conditioner
  fan, 311
  schematics, 311–315
  thermostat, 311
  troubleshooting, 315–318, 320–321
Ammeter, 112
  AC, 116
  extending the range, 113
  clamp-on tape, 114
Ampere, 5, 26
Asia Minor, 4
Atom, 8

B

Batteries, 62, 127
  maintenance, 134
  types, 127
    alkaline, 135
    nickel-cadmium cells, 135
    primary cell, 127
    secondary cell, 127
    specifications, 128
Bleeder resistor, 242

C

Capacitance, 181
Capacitive reactance, 167, 175
Capacitors, 167–179
  basic units, 171
  breakdown voltage, 171
  capacity, 170
  checking, 176
  electrolytics, 54, 173
    connecting, 174
    making an electrolytic, 174
  fixed, 53
  how it works, 169
  ratings, 240
  troubleshooting, 178
  types, 171
  variable, 55
  WVDC, 175
Cells, 128
  connecting, 132
  dry, 128
  in parallel, 72
Charge, 3
  electrical, 11
  negative charge, 3
  practical charge for, 15
Chips, 194
Circuits
  air conditioning, 307–321
  heating, 277–305
  refrigeration, 323–343
Circuit breakers, 60, 191
  boxes, 190
Circuit, electrical, 17
  connecting, 19
  closed, 31
Circuit protectors, types, 252
Cold anticipator, 270
Color code, 44
  gold and silver third bands, 45
Compounds, 7
Compressor motor relays, 243

Compressor motor relays (*contd.*)
  current type, 244
  potential type, 244
Compressor problems, 346
  PSC compressor, 350
  air-conditioner compressor, 351
Conductors, 16
Control devices, 257–275
Controllers, 208
  electronic, 208
Controls, 193
Coulomb, 5, 25
Current, 16
Currents in parallel circuits, 72
Currents in series circuits, 70

D

Delta, 141
  connection, 141
Differential amplifiers, 208
Diodes, 194
  PN junction, 195
  special purpose, 196

E

Electric heating system, 279
Electricity, 5
  future, 5
  sources, 23
  static, 3
Electromagnetism, 91
  using, 94
Electron, 3, 8
  controlling, 15
  flow, 16
  orbiting, 11
  outer shell, 12
  properties, 8
  valence, 12
Electronic charging meter, 122

Elements, 7
Energy, mechanical, 37
Energy saver switch, 339

F

Franklin, Benjamin, 5
Fuses, 60, 71
  types, 62

G

Galvani, 5
Gases, 6
Gas furnace operation, 278
Germanium, 193
GFCI, 250
Greeks, 3

H

Handling refrigerants, 345
Heat anticipator, 270
Heat pumps, 286–293
  combinations, 293
  requirements, 293
High-efficiency furnaces, 294–303
  electrical controls, 295
  schematics, 297–298, 301–303
  operation, 295
  sequence of operation, 296
Hot-gas defrost service, 102

I

Impedance, 181
Inductance, 148, 181
  changing, 148
  self-inductance, 148
  mutual, 151
Inductive circuit, 152, 162
  delay, 164

power in, 152
Inductive reactance, 152
   measuring X_L, 152
   uses for, 153
Inductors, 55, 147
   fixed, 56
   variable, 57
Instruments, electrical measuring, 107
Integrated circuits, 200
   dual in-line package (DIP), 201
   flat pack, 201
   multipin circular, 201
Ions, 12

J

Joule, 35

L

Ladder diagrams, 280, 319
Leak detector, 120
   electronic sight glass, 120
Limit switches, 271
Liquids, 6
Low voltage operation, 351

M

Magnesia, 4
Magnetic theory, 87
Magnets, 85
   electromagnets, 87, 92
   permanent, 85
   permeability, 88
   poles, 90
   shapes, 88
   temporary, 86
Magnetism, 3, 4, 85
   in a coil of wire, 92
Magnetite, 3
Main switches, 248

Manufacturer's diagrams, 282
Matter, 6
Measurement, units of, 25
Megger, 118
Meter movements, 107
   types, 108
      analog, 110
      digital, 119
Meters, 63
Molecules, 6–7
Motors, alternating current, 219–245
   capacitor-start, 228–232
   capacitor-start, capacitor-run, 236–237
   permanent split-capacitor, 232–236
   repulsion start, induction run, 227
   shaded-pole, 221
   split-phase, 223
   starting, 224
   three-phase, 238
Motor protectors, 242
Motor start relay, 257
Multimeter, 118

N

Neutrons, 8, 12

O

Objectives, Performance, 2, 22, 42, 66, 84, 126, 146, 166, 180, 192, 218, 246, 256, 276, 306, 322, 344
Oersted, 5
Ohm, Georg, 5
   law, 28
   law, examples, 29–34
   law, in other forms, 33
   unit of measurement, 26
   uses for Ohm's law, 34
Ohmmeter, 116

Ohmmeter (*contd.*)
  adjusting, 122
  using, 122

P

Parallax error, 112
Parallel circuit, 72
  characteristics, 72
  rules, 73
Performance Objectives, 2, 22, 42, 66,
    84, 126, 146, 166, 180, 192, 218,
    246, 276, 306, 322, 344
Plugs, air-conditioner, 308
Portable electric tools, 249
Power, 35, 182
  AC and DC, 127–145
  distributing, 185
  kilowatt-hour, 38
  meters, 185–189
  polyphase, 185
  waveforms, 185
Power factor, 184
Power relay, 257
Prefixes, 27
Protons, 8
Pressure control switches, 272

R

Rapid electrical diagnosis (RED), 338
Refrigerator, 323
  defrosting, 328
    accumulated compressor time,
      332
    cycle, 329
    hot-gas, 333
  start relay, 323
  thermostats, 323
Refrigerator-freezer combination, 324
  automatic defrost, 324
  manual defrost, 324

troubleshooting, 335–342
Resistance, 17
  in series, 67
Resistors, 36, 43
  fixed, 49
  fusible, 52
  potentiometer, 47
  rheostat, 47
  tapped, 50
  temperature compensating, 52
  types, 49
  variable, 47, 51
Run capacitors, 241

S

Safety, 247–255, 345
  electrical, 247
  precautions, 247
Schematic, 48
  reading, 64
SCR, 197
Semiconductor, 58, 193
  diodes, 58
  principles, 193
  solid state, 193
  defrost control, 202
  transistors, 59
Sensing, 202–208
  bridge circuit, 205
  humidity, 203
  thermistor, 202
  unbalanced bridge, 206
Sensors, 208
Series and parallel circuits, 67–83
Series circuit, 67
  rules, 70
Series-parallel circuits, 74
  determining current, 78
  determining resistance, 75
  determining voltage, 78
  resistance circuits, 75

INDEX 369

Siemens, 28
Silicon, 193
Sine wave, 136
    characteristics, 137
Single-phase current, 181
Solenoid, 85, 94, 262
    liquid line service, 100
    sucking effect, 95
    suction line service, 100
    valves, 98
        principles of operation, 98
Solids, 6
Solid state, 193
    compressor motor protection, 211
    control modules, 211–216
Start capacitors, 240–241
Switches, 19, 59
    push button, 309

T

Thales, 3
Thermal overload protectors, 260
Thermostats, 263, 310
    adjustments, 270
    bellows type, 263
    bimetallic type, 265
    heating and cooling, 266
        mercury contacts, 266
    microprocessor, 267
Three-phase current, 181
Time delay relays, 262
Tolerance, 45
Transformers, 57, 154
    audio, 158
    autotransformers, 159
    construction, 155

iron core type, 155
losses, 161
low-voltage control, 103
mutual inductance in, 154
power, 158
rf or radio frequency, 158
step-up, step-down, 157
types, 156
voltage transfer, 156
Troubleshooting, 345
    capacitor testing, 353, 360
    clamp-on meters, 356
    motor-protector relay testing, 353
    motor testing, 352
    procedure, 362
    testing centrifugal switches, 359
    using a system, 351
    using meters, 355
Transistors, 194, 197
    impedances, 199

V

Valves, applications, 99
Volt, 15, 26
    difference of potential, 15
Volta, 5
Voltages in series, 69
Voltmeter, 115

W

Water tower controls, 273
Wiring, 282
    field wiring, 282–284
    low-voltage wiring, 284
Wye, 141
    connection, 143